FARM ANIMALS AND THEIR MANAGEMENT

FARM ANIMALS AND THEIR MANAGEMENT

By

JAMES A.S. WATSON

WATTIE J. MILLS

K.L. BLAXTER

C.C. THIEL

SSPH

SATISH SERIAL PUBLISHING HOUSE

403, Express Tower, Commercial Complex
Azadpur, Delhi-110033 (India)
Phone : 011-27672852, Fax : 91-11-27672046
e-mail : info@satishserial.com, hkjain1975@yahoo.com
Website : www.satishserial.com

Published by :

SATISH SERIAL PUBLISHING HOUSE
403, Express Tower, Commercial Complex, Azadpur, Delhi-110033 (INDIA)
Phone : 011-27672852 Fax : 91-11-27672046
E-mail : info@satishserial.com, hkjain1975@yahoo.com

Reprint 2020

ISBN 81-90228-95-1

Composed, Designed & Printed in India

Preface

This book was originally intended as a text for students, and its primary purpose has not changed; it has, however, been increasingly used by practising farmers and in order that it may serve their needs also, the authors have been at pains to avoid the use, without explanation, of such scientific terms as have not yet become part of our farming vocabulary.

Because of the lamented death of James Anthony More, and also by reason of ever-increasing specialization in Agricultural Science, it has been necessary to seek the collaboration of several friends and colleagues.

Blaxter has revised the chapter on Animal Nutrition and Dr. C.C. Thiel the section on Dairy Equipment. The late Dr. D.J.G. Black, just before his death, re-edited the chapter on Poultry. Miss Eileen Handlen has again provided the Index.

A great part of the book has been rewritten and much new material has been added.

The pace of progress in farm science and technology is now such that any sizable book must inevitably be out of date, in some respect, before it leaves the press. In the present case the most important failing is with respect to various noteworthy developments in the field of farm animals and their management, sheep dips, etc.) which occurred in the few months preceding publication.

James A.S. Watson

Illustrations

Contents

CHAPTER I

FARM LIVESTOCK

GENETICS: ANIMAL BREEDING
ANIMAL REPRODUCTION

IN animal reproduction the embryo develops from a fertilized egg, which is the product of the fusion of an egg cell produced by the female with a sperm cell produced by the male. In some species the fertilized egg divides, giving rise to two or more embryos, and one-egg or so-called "identical" twins occur occasionally in cattle, though not in other domesticated animals. A large majority of twins in cattle are "ordinary" or two-egg twins.

The primary sexual organs of the female are the ovaries, and the shedding of the egg from the ovary may be taken as the start of the reproductive process. The ovary, apart from producing eggs, gives rise to internal secretions which control the restrous cycle. At or about the time of shedding of the egg occurs the period of restrus ("season" or "heat") when mating takes place. The ruptured follicle, from which the egg has been shed, forms the "yellow body" *(corpus luteum),* and so long as this persists restrus does not recur. Normally, however, in the non-pregnant animal, the yellow body disappears after a fairly definite period of time, when another follicle ruptures and the animal comes "in season" again.

In the normal healthy cow successive heats recur at fairly regular intervals of about twenty days throughout the year, though in winter the period of heat may be very short. In the mare the interval is about twenty-one days, but restrus does not recur regularly during winter. In the sow the first heat occurs shortly after weaning, and in the non-pregnant animal recurs throughout the year at intervals of twenty-one or twenty-two days. In the ewe, by contrast with other species, restrus is more or less confined to the autumn, the main controlling factor being length of day;

thus ewes can be brought into restrus during summer by subjecting them to nightly periods of darkness of twelve more hours. As would be expected, sheep kept in tropical regions show no breeding seasonality, and the length of the breeding season falls with increasing latitude. The peak of the breeding season, however, is usually somewhat before the shortest day. Again, the length of the breeding season varies from one breed to another. At the one extreme are the wild species, which ordinarily have only one or two heats per year; at the other is the Dorset Horn, which may have twelve or more. Most wild birds produce only a single clutch, or only two, during the year, but some, such as the wood pigeon, may have three or four. Domestic poultry may continue in full lay for many months; here, as in the case of sheep, day length has an important influence on ovulation.

Cestrus during pregnancy is abnormal, and it is ordinarily assumed that a female that has been mated, and has ceased to " come in season," is pregnant. But, on the one hand, it is not unknown for a pregnant female to show all the symptoms of restrus, and to take the male. On the other hand, the *corpus luteum* sometimes persists in the non-pregnant animal much beyond the usual time-the case of so-called "false pregnancy."

Single ovulations are normal in the cow and mare, and it is doubtful whether twinning in these species is desirable under ordinary farm conditions, twins being, on average, less efficient than singles. In the ewe, double and treble ovulations are common, at least in certain breeds, if the level of nutrition and general well-being is high. Except under mountain conditions, a proportion of twins is desirable. In the sow, the number of eggs shed at each heat is generally of the order of twenty, but a considerable number of the embryos ordinarily degenerate at an early stage of development. Litters of twenty and over are, however, by no means very rare.

The average gestation period of the mare is about eleven months (340 days), in the cow nine months (280 days), in the ewe five months (150 days), and in the sow sixteen weeks (112 days). There is, however, a considerable range of variation.

Parturition, of course, normally coincides with the commencement of lactation, and the distention of the udder with colostrum is one of the signs that parturition is imminent. But animals on a high level of nutrition, and especially those that have been bred for high milk production, may begin to " run out" some time before giving birth. At the other extreme females of the specialized beef types, especially if they are in very low or over fat condition and have troublesome calvings, may have little or no milk.

After a female has given birth there is a longer or shorter interval before the recurrence of restrus. In the mare this interval is short-commonly four to seven days. In the cow it is longer and less uniform, but commonly between one and two months. The sow rarely comes on heat while she is nursing a litter, but usually does so on the second or third day after the litter has been weaned.

The age at first oestrus depends to some extent on the breeding of the animal and the level of nutrition. Heifer calves may come on heat as early as six months of age and gilts at about four months. With spring-born ewe lambs some 70 or 80 per cent., on average, come on heat during the shortest-day period of their first year, the percentage being highest in those born early and well-nourished throughout their lives. With late-born lambs reared under hill conditions the proportion is low, but it is still necessary to take precautions to prevent mating, which is very undesirable under mountain conditions.

Artificial Insemination: Since a single ejaculation of semen normally contains many millions of spermatozoa, it is theoretically possible for a single male to beget vast numbers of progeny. In practice it is possible, by artificial insemination, to multiply the normal number from five to twenty-five times, and thus, to exploit the value of a selected male. There is no great difficulty in collecting semen from the bull, ram, stallion, or boar. If the semen is diluted with a nutrient solution and is kept at an appropriate low temperature it retains its vitality, in the case of the bovine, for several days. Moreover, under deep-freeze conditions (-79°C) vitality is maintained for a period of years. The advantages of deep-freezing are four, viz. (1) The semen of a particular bull can be made available, in practice, at any time that it may be required (instead of only three or four days per week under cool storage); (2) a stock of semen can be retained after the death of the male animal which has produced it; (3) when there is a limited breeding season *(e.g.* in New Zealand dairy herds) collection of semen can proceed throughout the year, so that only a relatively small number of males need be kept; and (4) the insemination service is less liable to interruption by the occurrence of epidemic disease in the area served.

THE MECHANISM OF HEREDITY

The egg is a relatively large cell, containing a mass of protoplasm and yolk, together with a small nucleus-a clearly differentiated region with a characteristic structure. The sperm is a very minute motile cell with no yolk and very little cytoplasm *(i.e.* extra-nuclear protoplasm), but containing a nucleus that is similar in size and appearance to that of the egg. There is ample evidence that the bearers of the hereditary characters must be sought for in the nuclei of the germ cells. For one thing, the only portions of the egg and sperm that are comparable are their nuclei, and we know that, broadly speaking, the male parent has just as much influence on the characterization of the progeny as has the female. The nucleus of a cell that is in normal growth, or is carrying out its ordinary functions, contains a complex' network of threads of a readily stainable substance called chromatin. When, however, cell division is about to occur, the chromatin threads condense, thicken, and assume the form of a number of generally rod-like bodies called *chromosomes.* These chromosomes are constant in number for all the cells of the individual and (with the significant exception mentioned below) for all the individuals of a race. Moreover, it can frequently be shown that each chromosome has a definite kind of individuality, for whenever particular members of a set can be recognized by size or sbape, a group from one cell can be matched, member for member, with that from another. A further important fact is that the chromosomes occur in the body cells in pairs, and all the evidence points to conclusion that of each pair one member is derived from the male, the other from the female parent. In the ordinary type of cell division (Fig. 1) each chromosome splits longitudinally into two like halves; the latter are drawn apart to opposite, poles of the dividing cell, the nuclear wall breaks down and two nuclei are formed, each containing a complete set of the divided chromosomes.

In the division that gives rise to the reproductive bodies the ova or the sperms-the process is essentially different from that just described. The condensation of the chromatin into chromosomes takes place much more slowly, and when the chromosomes finally appear, they are found to be united in pairs. The long preparatory stage is believed to have to do with an exchange of like parts between the paternal and maternal members of each chromosome pair. In any case the two members of each chromosome pair now separate, a further division follows immediately, and four nuclei are produced. The essential point is that the nucleus of each cell contains only half the number of chromosomes that is characteristic of the body cells of the species.

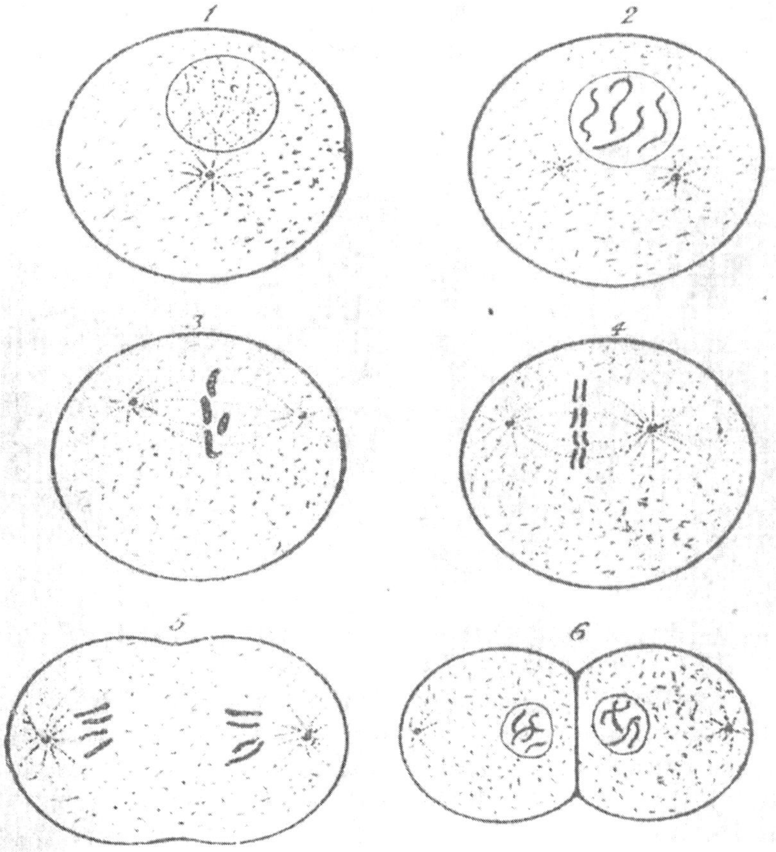

FIG. 1: Ordinary Cell Division

Of the four cells, each, in the case of the male, becomes a functional sperm (see Fig. 2). In the case of the female, one becomes the functional egg while the others are cast off as the so-called polar bodies, which can be looked upon as abortive eggs. When fertilization occurs the sperm nucleus fuses with that of the egg and the original (or double) number of chromosomes is restored.

A further interesting point is that there is frequently a characteristic difference between the set of chromosomes possessed by the male and by the female respectively. The one sex has *a* complete double set of chromosomes, the two members of each pair being like. The other sex has either one odd chromosome or else an unlike pair. Thus, in the American Fruit Fly, *Drosophila melanogaster* (see Fig. 3), the female has four like pairs of chromosomes, and when reduction occurs, produces, so far as the chromosome-content is concerned, only one kind of egg. In the male, three pairs of chromosomes and one member of the remaining pair can be matched in the female set, but the second member of the last pair is recognizable from the others by its shape, which is sharply bent. The male, thus produces two sorts of sperm-cells in equal numbers, half containing the straight chromosome (or X-chromosome as it has been called) and half the bent or V-chromosome. It follows that eggs fertilized by the first sort will produce females, since they will possess the constitution symbolized by the formula XX; those fertilized by the second will develop into males, since they will have the constitution XY.

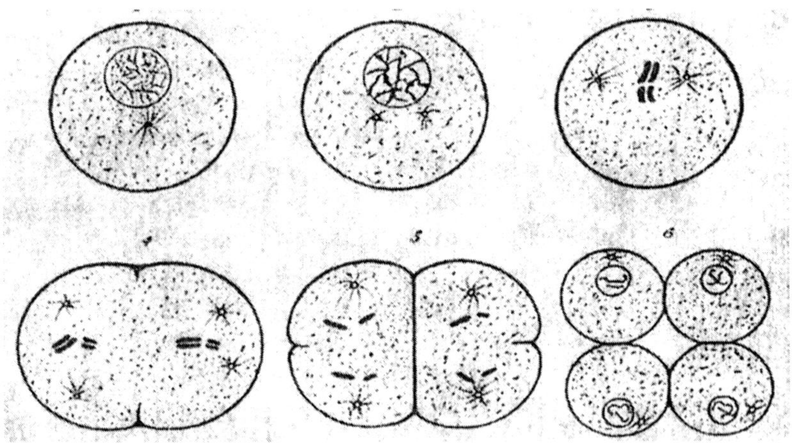

FIG. 2: The Reducing Division-Formation of reproductive cells in the male

The chromosomes, thus provide a simple mechanism for the determination of sex. Apart from this, the facts just quoted are important in connection with what is called sex-linked inheritance, a peculiar phenomenon that arises from the circumstance that factors for other characters than sex are carried by the sex chromosomes.

The particular arrangement of the sex chromosomes just described is not universal. It is reversed, for example. In fowls, where the male is *homogametic* (producing but one sort of sperm so far as chromosome-content is concerned), while the hen is **heterogametic**, giving rise to two sorts of ova, male-producing and female-producing respectively.

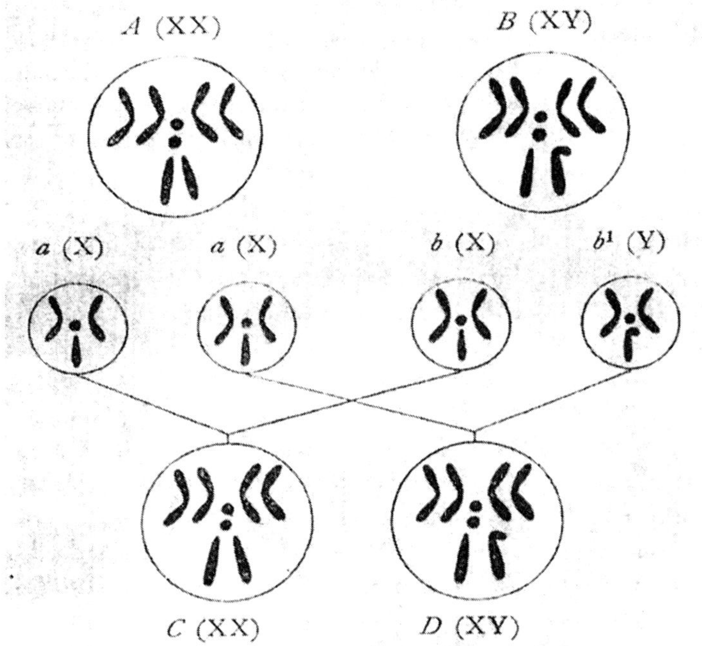

FIG. 3: The Chromosomes in *Drosophila melanogaster*

A, female; B, male; a, a, ova; b, b^1, sperms; C, female zygote formed by union of sperm b with egg a; D male zygote formed by union of egg a with sperm b^1

MENDEL'S FIRST LAW

Such being the mechanism and such the general theory of heredity, let us now turn to some of the observed facts of inheritance. In 1865, Gregor Mendel, as a result of some crossing experiment with garden peas, was led to make an important generalization regarding the manner in which parental characters are distributed among the progeny. Varieties of peas show a considerable number of definite character differences-e.g. some varieties are tall, others dwarf, some have round seeds, others wrinkled, some green seeds, others yellow, and so on. Moreover, the plants are self-fertilizing, and varieties, as a rule, breed true to their own particular characters. Mendel's method was to cross two types showing some definite character difference; if the two varieties showed more than one such difference he considered each difference without reference to the others. He reared the hybrid plants and allowed them to become self-fertilized; in the next or second hybird generation he classified the individual plants with reference to the original parental character, and counted the members of each group. In one experiment, for example, Mendel crossed an ordinary tall variety with one of the well-known dwarf sorts, both varieties being known to breed true to their respective type. The first-cross plants were tall like the tall parent-indeed, they somewhat exceeded the latter in height. These hybrid talls did not, however, breed true. On the contrary, when allowed to become self-fertilized, they produced tall and dwarf progeny in the proportion (roughly) of three of the former to one of the latter. Mendel's actual figures were 787 tall and 277 dwarf, or 73 and 27 per cent, respectively. When these second-cross plants were tested individually (by making separate sowings of the seeds of each), it was found that all the dwarfs bred true. Of the tall second-cross plants a proportion, about one-third, were found to produce nothing but tall progeny, while the remainder gave tall and dwarfs in the same ratio as did the first cross. These results may be summarized in the following scheme:

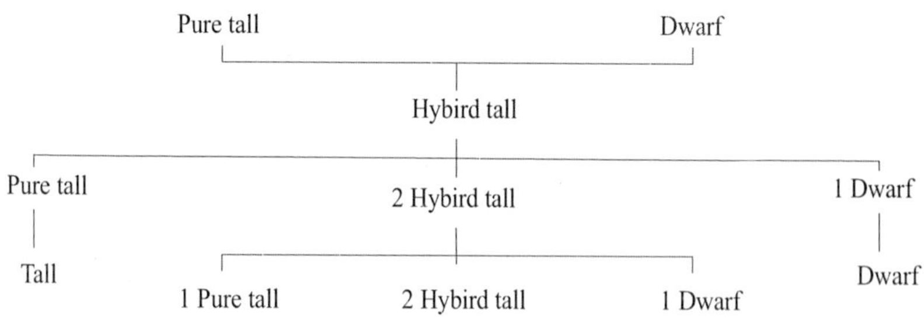

The fact that the first cross resembled the tall parent and showed no sign of the dwarf character was expressed by Mendel by saying that "tallness" was *dominant,* and "dwarfness" *recessive.* Dominance, however, is by no means universal; sometimes a hybird shows a condition somewhat intermediate between those of its parents. For instance, if a red-flowered variety of *Mirabilis jalapa* (Marvel of Peru) is crossed with a white, the hybird bears pink flowers. If the hybird is selfed and a second hybird generation is reared, the ratio obtained is 25 per cent pink, and 25 per cent white. The reds and whites so obtained breed true, while the pinks of the second generation give the three types in the same ratio

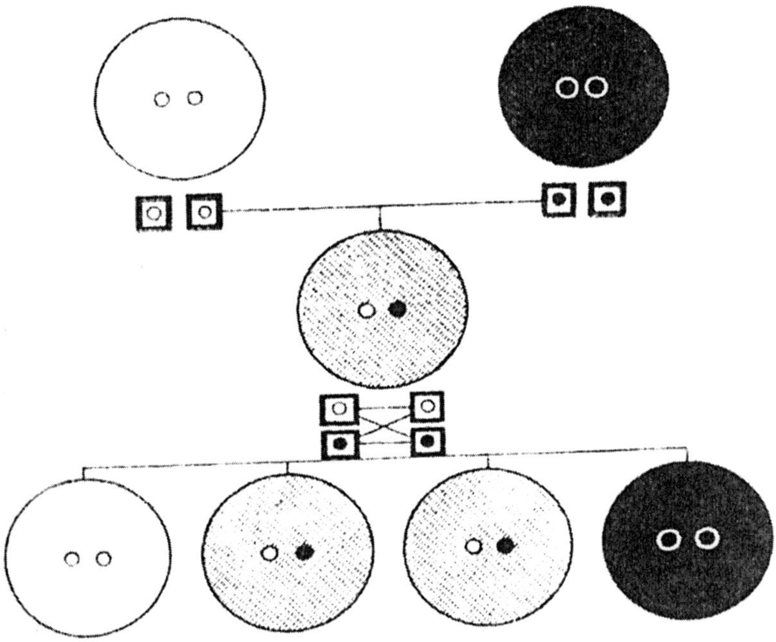

FIG. 4: Cross between Red and White Varieties of *Mirabilis jalapa*

The large circles represent the parent, first hybird, and second hybird generation, and the small circles the genes. The squares represent the reproductive cells.

as did their parents. Thus, whether or not there is dominance in a particular case, the general result is the same. When hybrids are self-fertilized or interbred they produce, with regard to any particular character, three types of progeny; the original parental types are each reproduced in one-fourth of the second hybrid generation, while the remaining half are hybrids like the first cross.

Mendel not only discovered this general rule of inheritance but he proposed an explanation of the facts. His hypothesis still stands. Subsequent research has brought out much evidence in its support, and has served to develop rather than to modify the original theory. Mendel knew nothing of chromosomes, but the facts regarding their behaviour fit in remarkably well with the phenomena that he observed, so that to-day we may best state Mendel's hypothesis in terms of the chromosomes. We assume, then, that a "factor" or "gene" controlling, say, the flower colour of the *Mirabilis,* is borne in one of the chromosomes of the species. A red-flowered plant will have a red-producing gene in both members of the chromosome pair concerned, and since each egg or pollen cell that it produces will receive one member of the pair, only one type of reproductive cell will be produced. The white-flowered plant will contain "white" genes in the corresponding region of its corresponding chromosome pair, and all its reproductive cells will carry the white factor. When the two varieties are crossed it is obvious' that the hybrid will have a pair of chromosomes that are unlike with regard to the gene for flower colour, one having the red factor and the other the white. What visible result will be produced by such a combination is not capable of prediction, but in the present case we saw that the hybrid had pink flowers. Now the essential point of the theory is that the two genes do not intermingle but remain throughout the life of the plant as separate and distinct entities. When the reproductive cells are formed, and the two members of the chromosome pair separate, half the ova and pollen grains will receive a chromosome containing the "red" gene, while the other half will get one containing the "white." The pink plant produces, then, reproductive cells that carry not a factor for the pink character but that for either the original red or the original white, and the two classes occur in equal numbers. When the hybrid pink plants are selfed, fertilization of either type of egg by either type of pollen grain occurs as a matter of "pure chance," so that the following four combinations are equally probable, viz. : " white" pollen grain with "white" egg; "white" pollen grain with "red" egg; "red" pollen grain with "white" egg; "red" I pollen grain with "red" egg. The first combination will produce a white-flowered plant, the last a red, and 'the second and third pinks that are not distinguishable, since it is immaterial whether the red is introduced from the male or the female side. Thus, is produced, it is suggested, the ratio of 1 red: 2 pink: 1 white.

The actual numbers obtained in Mendelian experiments are subject to chance variations, and do not, except by chance, give exactly the "expected" ratio. The numbers can be predicted only in so far that we can state the most probable result, just as we can in the case of say, a series of tossing of a coin. Thus, if we make a hundred throws, the most probable single result is fifty heads and fifty tails, but the probability of obtaining fifty-one or forty-nine heads is only very slightly less, and so on through the whole series of possibilities. When, however, large numbers are dealt with, the ratio should approach very closely to theoretical expectation, and this has been the case in Mendelian experiments. For instance, one of Mendel's original experiments, dealing with cotyledon colour in peas, has been fourteen times repeated by different experimenters, and a total of 203,500 second hybrid individuals have been bred. Yellow colour is completely dominant to green, so that the expectation was 3 yellow, 1 green; or altogether 152,625 yellow and 50,875 green. The totals actually obtained were 152,824 yellow and 50,676 green-a ratio of 75.1 per cent, to 24.9 per cent.

If the hypothesis that has been outlined above is adequate, it ought to enable us to predict the results of a mating between any two of the three possible types. For example, if we cross the pink (heterozygous[1]) *Mirabilis* back to the parent red form we should expect the following two combinations in equal numbers: "white" pollen and "red" egg cell, giving pink; "red" pollen and "red" egg cell, giving red. Similarly, a cross of pink and white should give us pink and white progeny in equal numbers. Results of this sort have been regularly obtained.

Mendel's first law may, therefore, be stated thus: When a dissimilar pair of factors are brought together in crossing they separate again when the hybrid produces its reproductive cells. The separation is clean and complete. The hybrid condition cannot be represented in the sperm or in the unfertilized egg it can only be produced in the fertilized egg, and then by the union of an egg and a sperm that are dissimilar with regard to the factor in question.

MENDEL'S SECOND LAW

Mendel's second generalization was to the effect that when two or more character pairs are considered simultaneously, each pair behaves without reference to the others. Since Mendel's time, indeed, a very

[1] Heterozygous, since it received from its parents dissimilar factors for the character nw being discussed, a "red" factor from the one and a "white" from the other.

important group of exceptions to this law has been discovered, and the evidence leads to the conclusion that these exceptions arise in cases where two pairs of genes are contained in one and the same pair of chromosomes. With this reservation, Mendel's second law is true. Take, for example, the results obtained ion crossing a tall purple-flowered variety of pea with a dwarf white-flowered sort. The tall condition is dominant to the dwarf and coloured is dominant to white. The first cross is, therefore, tall purple. If these plants are selfed and the succeeding generation reared we get, of course, the ratios of 3 tall: 1 dwarf and 3 purple: 1 white. But the two dominants do not necessarily go together. On the contrary, the tall group will consist of purples and whites in the ratio of 3 : 1; the whites will consist of 75 per cent, tall, 25 per cent, dwarfs, and so on Hence, the combined ratio is obtained by algebraic multiplication of (3 T + 1 D) x (3 P + 1 W), giving 9 tall purple: 3 tall white: 3 dwarf purple: 1 dwarf white. It is obvious that very complex ratios will be produced when three or more character pairs are considered simultaneously. It was this complexity, apparent rather than real, that baffled all Mendel's predecessors, and it was Mendel's plan of dealing with one thing at a time that brought out simplicity and order from seeming chaos.

One or two somewhat odd cases must be described in which, though inheritance is strictly Mendelian, the results are complicated by reason of the effects produced by particular factors. One of the earliest discovered cases had to do with the inheritance of yellow coat-colour in mice. The first peculiarity of the case is that yellow mice never breed true; in matings of yellows with yellows the young are about two-thirds yellow and one third of other colours. Yellows are never produced from matings of two non-yellows. If yellow is mated with non-yellow the young consist of equal numbers of yellows and non-yellows. These facts indicate that yellow is dominant, but that all yellow mice are heterozygous for the yellow condition. If two such are mated, the expectation would be 1 pure yellow: 2 hybrid yellow: 1 non-yellow, and the only explanation of the facts is that the pure or homozygous yellow is incapable of life. This has found confirmation in the fact that in yellow x yellow matings a proportion of the embryos die in the early stages of development. Why the homozygous yellow should be incapable of life beyond this early stage is not known, but it is interesting to note that even the hybrid yellows are not quite normal, being distinctly less hardy than non-yellows, and liable to become excessively fat. A further interesting fact is that many analogous *lethal factors,* leading to structural abnormalities or functional derangement of such an order as to render the individual incapable of living, have been discovered in several species, including cattle and sheep.

A very striking result of hybridization was obtained in the course of some of the earlier experiments with sweet-peas. Two white varieties gave a purple first cross. This occurrence can properly be classed as a reversion, since the wild parent of our cultivated forms is purple-flowered. These crosses, when left to become self-fertilized, gave progeny in the ratio of 9 purple 7 white. In other crosses between two whites, other colours than purple were obtained in the first cross, and sometimes the second hybrid generation contained a mixture of coloured types, the proportion of coloured to white remaining, however, that of 9 : 7. Moreover, the whites, when self-fertilized, always bred true; some of the coloured types produced only coloured progeny, while others gave one-fourth and still others seven-sixteenths of whites. The explanation of this case is that colour is produced only in the presence, simultaneously, of two dominant factors. One of the white varieties contains the first of these factors but lacks the second; the other has the second but lacks the first. The first hybrid, thus contains both dominants and is coloured. In the second hybrid generation we get as usual a 9 : 3 : 3 : 1 ratio, the last three classes lacking respectively one or other or both dominants, and hence being white. Many implications of this theory have been tested and the expected results obtained. The occurrence of different colours in the second hybrid generation is explained by the fact that white sweet-peas may carry various factors capable of modifying colour although, in the absence of the colour factors themselves, these can have no visible effects.

A long series of similar cases might be described; none of them, however, constitutes an exception to Mendel's two laws. The complications arise out of the visible effects of the factors, and the factors themselves behave in the normal way.

LINKAGE

It was not until some ten years had been spent in active Mendelian investigation that any new principle was discovered.

The first cases that failed to conform to Mendel's second generalization were found in sweet-peas. In one of these cases a variety with purple flowers and an ordinary erect standard petal was crossed with another having red flowers and a hooded standard. It was previously known that purple behaved as a simple dominant to red, and that the erect standard was a simple dominant to the hooded. The first hybrids were purple erect, according to expectation. Had the case been normal the second hybrid generation should have consisted of the four possible combinations in the proportions 9 purple erect: 3 purple hooded: 3 red erect: 1 red hooded. The actual numbers obtained were 2036 purple erect: 12 purple hooded: 10 red erect: 654 red hooded. The numbers show that the two dominant factors on the one hand and the two recessives on the other have somehow "stuck together" in a great majority of cases. The numbers, in fact, strongly suggest the simple ratio of 3 : 1-three double dominants to one double recessive. But the other two classes, each containing one dominant only, do occur, though relatively very rarely.

When the cross was made between two parent plants with the same characters in the opposite combinations, the result was quite different. Purple hooded crossed with red erect again gave a purple erect first cross. In this case, however, the second hybrid generation consisted of 2969 purple erect: 1369 purple hooded: 1441 red erect. The fourth possible class, *i.e.* the double-recessive red hooded, did not occur at all. Here, again, the results suggest a simple monohybrid ratio-viz. one-fourth like each of the original parents and one-half like the first cross. It is obvious from the second experiment that the case is not one of attraction between any two characters as such. The point is, however, that the characters are combined in the original parents, these combinations tend to persist and to re-appear in the second hybrid generation. The old combinations are broken and new combinations made only in a small minority of cases.

This phenomenon of *linkage,* as it is called, arises whenever two pairs of genes are carried in one and the same pair of chromosomes. For a full explanation the reader is referred to any standard work on genetics.

In the examples quoted above there are *pairs* of Mendelian factors e.g. tall and dwarf in peas, red and white *in Mirabilis.* But many cases are known *of multiple allelomorphs.* Bay, black, and chestnut colour in horses constitute such a series-i.e. a bay horse may carry the factor for black or chestnut (but not both) while a black may carry chestnut. The dominant grey factor is separately inherited; only in its absence can a horse be of any other colour.

QUANTITATIVE INHERITANCE

The vast majority of economic characters of farm animals size, milk yield, egg production, quality of wool, etc. in so far as they are influenced by heredity, depend on a large number of genes, whose individual effects cannot be isolated. Such are known as polygenic characters. Moreover, as is well known, such characters are influenced not only by heredity but also by environmental conditions. Thus, a Friesian cow (apart from the possibility that both her parents carried the recessive red factor) will be black and white whatever the conditions under which she is reared. But the expression of her genetic milk producing capacity will vary with age, disease, level of nutrition, the degree of skill exercised in management and milking, and many other circumstances. Under extremely unfavourable conditions her actual milk yield will bear very little relation to her genetic make-up.

Hence' a useful distinction can be drawn between the *genotype* and the *phenotype.* The genotype is the result of the random union of gametes produced by Mendelian segregation within the restrictions caused by linkage. The phenotype is determined partly by genetic and partly by environmental forces, and by the interaction between the two.

The relative influences of heredity and environment vary as between one character and another. For example, in the dairy cow the butter-fat content of the milk is much more a matter of heredity, and much less a matter of environment, than milk yield. The proportion of phenotypic variation of a particular quality that is due to heredity is termed its *heritability* and is commonly expressed either as a decimal fraction or as a percentage of the total variation. Thus, the heritability of the butter-fat content of milk, within a given herd, is commonly about 0.4 (40 per cent) while that of milk yield is much less-20 or 25 per cent. Thus, it is easier to breed for butter-fat than for milk yield.

SEX-LINKED INHERITANCE

A complication is introduced in the transmission of certain characters owing to the fact that their factors are carried by the sex chromosomes. In poultry, for example, the male has a like pair of sex chromosomes while the hen has only one functional or gene-bearing sex chromosome. The result of this condition of affairs is illustrated by the behaviour of the barred pattern of the Barred Rock, which is dominant to black and sex-linked. If a pure-breeding barred cock be mated to black hens, the progeny are all barred. In the second generation the males are all barred and the females half and half as to colour. When the cross is made in the opposite direction-black cock with barred hens-the male chicks are barred and the females black (criss-cross inheritance). In the second hybrid generation we get the four classes-black males, barred males, black females, and barred females in equal numbers. The mechanism is illustrated in Figs. 31 and 32.

The explanation of all sex-linked cases is essentially the same. One of the sexes-the female in the case of the domesticated fowl-has only one functional or gene-bearing sex chromosome. This it obtains in every case from its parent of the opposite sex. The other'sex has a pair of gene-bearing chromosomes, one derived from either parent, and therefore, the normal complement-a double set-of the sex-linked genes. Its visible characters, in the event of the genes being different, are determined by dominance.

Sex-linkage has an important application in poultry breeding (see Chapter VIII), since with certain crosses the sexes of the chicks can be distinguished immediately on hatching.

We may now summarize our general conception of heredity. The hereditary characters of an animal are determined by a double set of factors or genes-one set derived from either parent. The sole exceptions to this rule are the sex-linked characters which, in one or other of the sexes, are represented by single genes only. Ordinarily each visible character is the resultant effect produced by a pair of genes, whether the two members of the pair be like or unlike; sometimes the effect of one gene is so strong as completely to mask that of its fellow (dominance). When the individual produces its reproductive cells and the chromosomes are reduced to half their original number, each egg or sperm cell receives either the one or the other member of each pair of genesthat is, it receives a single set. The mechanism controlling distribution of the genes between the reproductive cell is such that certain pairs assort quite independently of others. This happens when they are situated in different chromosomes. There is, however, a tendency, definite and measurable in particular cases, for certain genes to stick together in pre-existing combinations (linkage). This, it is believed, is due to their occurrence in one and the same chromosome. Linkage, however, is normally only partial, so that the chromosome cannot be regarded as a permanent or immutable assemblage of genes.

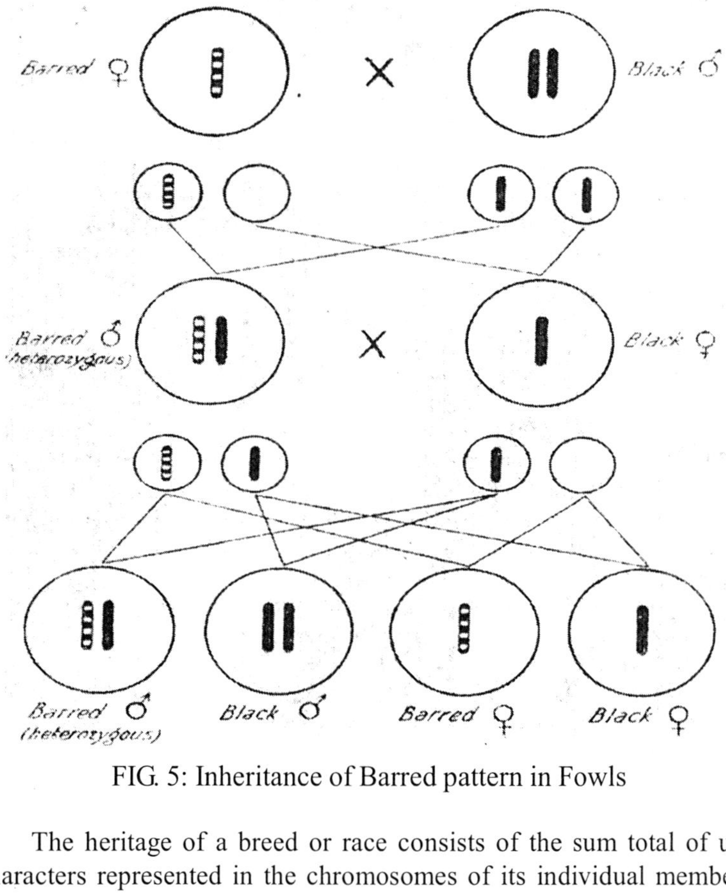

FIG. 5: Inheritance of Barred pattern in Fowls

The heritage of a breed or race consists of the sum total of unit characters represented in the chromosomes of its individual members. The great bulk-of observed variation between individuals of the same race (in so far as it is not due to conditions of nurture) is due to the process of biparental inheritance. This involves a combination, in each individual, of two sets of genes, one derived from each parent, and a reshuffling of these before they are handed on, in the germ-cells, to the next generation. A race will breed true to all its characters only when each and every pair of genes consists of two identical members, *i.e.* when the maternal set of genes is exactly matched by the paternal.

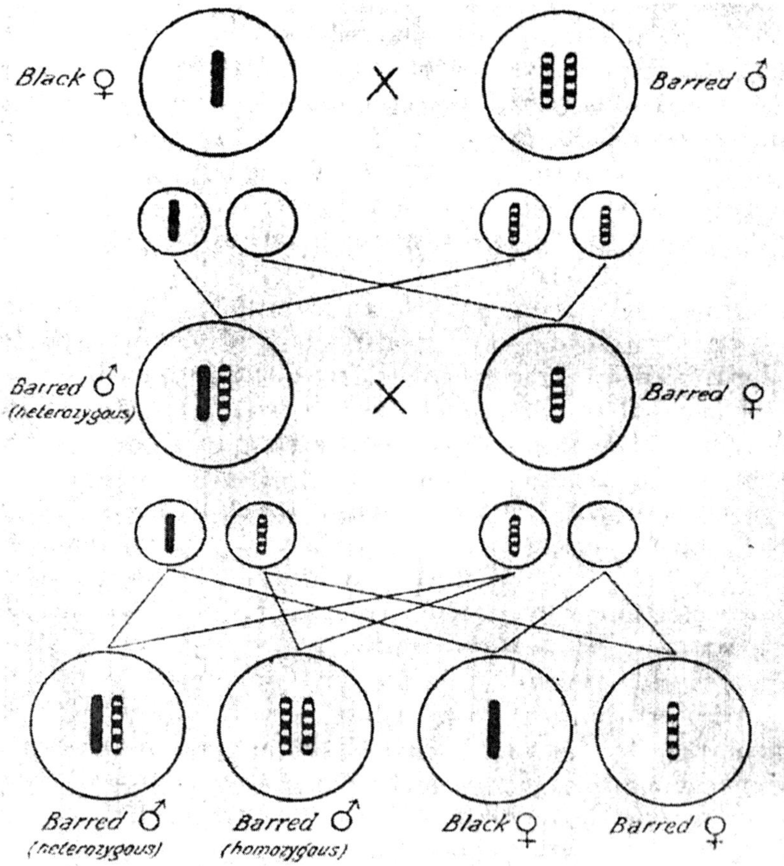

PURE LINES

It is worth while to look into this hypothesis and see whether it will stand the test of experiment. The effect of biparental reproduction is to maintain a large number of characters in the heterozygous condition, and thus to produce a large amount of variation. In fact, the whole significance of sexual reproduction seems to be that it makes possible a great series of character combinations and thus gives, the race as a whole the power to adapt itself to changing conditions. Among higher animals, therefore, we cannot expect to find true-breeding strains unless we employ artificial means to produce them. Many plants, however, have abandoned biparental reproduction, resorting either to vegetative methods or else to self-fertilization. In either case we get rid of the complexities produced by the reshuffling of genes in each successive generation, and therefore, we should expect to find fixed, true-breeding strains. Such variations as occur within the strain must be due to causes other than heredity-i.*e.* to environmental causes-and should not be transmitted. This theorem has been put to the test in many experiments, and the result, in very case, have been in agreement with it. For example, if we take a single barley plant descended as normally it would be, from self-fertilized ancestors, it will produce a group of progeny having strictly the same hereditary characters as itself. Such a group of progeny, constituting what is called a *pure line,* is fixed, because all its members have the same set of characters in the homozygous condition. Selection within a pure line accomplishes nothing, because selection can produce results only where hereditary differences exist. Cross-fertilization between plants of the same pure line is likewise without effect, for it amounts to nothing more than a combination of like with like. We are, therefore, justified in concluding that, as a rule, the individual genes remain constant from one generation to the next, and that the bulk of observed variation among the members of a breed or race is due to re-assortment of unit characters and not to any change in the unit characters themselves. This is important from the practical point of view, for it implies that there is a definite limit to the possibilities of selection. When we have obtained a certain combination of characters, all in the homozygous condition, we get no more variations of hereditary value and hence, further selection is a ploughing of the sands.

Such is true in the main; the units that compose the hereditary material are of definite constitution, and normally they remain constant. But occasionally, for reasons that are not understood, the genes do change. And when change occurs it is of the nature of a definite step; the new character is fully as constant as the old and behaves as a Mendelian alternative to it. Such sudden and definite changes in the hereditary materials are known as *mutations;* they have been seen to occur in most species of plants and animals that have been kept under continuous and close observation; they are not necessarily connected with sexual reproduction-vegetative sports being relatively common. Sometimes, they are obvious and startling, more often they are to be detected only upon close scrutiny, and it is certain that many have passed unobserved or have not been distinguished from the ordinary sort of variation due to recombination of factors. Mutations can be produced in plants by treatment with certain chemical substances and by certain ray treatments, but their direction cannot be controlled and nothing of economic value has so far been produced by artificial means. All that the breeder can do is to watch for their occurrence and to perpetuate such as happen to prove of use.

Mutations, as has been said, occur only rarely. In small laboratory animals, which can be bred in large numbers, it has been possible to estimate their frequency. In the fruit fly *(Drosophila)* the total mutation rate is about 0.7 per cent, *i.e.* seven out of every thousand sperms or eggs carry a mutation in one or other of their genes. The total number of genes in *Drosophila* is probably of the order of five or six thousand, so that, on average, the chance of a mutation in a particular gene is slightly more than one in a million. Moreover, of the many mutations that have been observed in farm animals, very few have been of economic value. Yet in the mass, the mutations that have resulted *e.g.* in the evolution of the superfine merino fleece from the hairy coat of the original wild sheep-have been of immense importance.

LIVESTOCK IMPROVEMENT

The business of the constructive breeder' who sets out to improve his herd rather than merely to multiply it is to build up a certain combination of valuable hereditary characters-that is, to produce a good type of animal that will "breed true" to its good qualities. Obviously the first essential is a definite object, a clearly conceived ideal type. Moreover, this ideal should be no mere arbitrary standard of "points" but should be laid down with regard mainly to economic considerations-that is to say, our ideal animal should be that which will convert its food into meat or wool, milk, or work, as the case may be, with a maximum of efficiency.

In certain cases it is easy to devise a simple and direct test of economic worth. It is so, for example, in the case of milk or of wool production, because it is possible in both instances to determine both the yield and quality of the product and to take such determinations as a basis for selection. The usefulness of a work animal is more difficult to assess, because no final judgment can be reached until the end of its working life. With meat-producing animals, again, no complete or definite information can be obtained on such important points as the quality of the meat or the proportion of carcass to offal, until the animal has been slaughtered. In general, however, any direct test is preferable to a mere inference based on the animal's outward appearance, and there is an increasing tendency to apply such direct tests, wherever possible. With pigs, for example, pro-lificness can be recorded and the milking capacity of a sow can, for practical purposes, be sufficiently measured by determining the live-weight increments of her piglings during the period when they are solely dependent on their dam's milk. The efficiency of the animal in converting food into meat can be expressed as a ratio between its weight increment and its food consumption, and even on such a matter as the quality of the meat of a breeding animal, inferences can be drawn with some degree of certainty if a few of its litter brethren or its progeny be fattened and slaughtered.

It is important to note that improvement implies the closer adaptation of the animal to the environment in which it is expected to live, and that a breed or strain that has been improved in relation to one type of environment may have been made less well-adapted to another. For example, the type of sheep that has been evolved for lowland conditions, which grows and fattens rapidly on abundant and nutritious food, lacks the agility to find its food on a mountain pasture, and its requirements for minerals to make bone, and for protein to make flesh, are so high that it cannot collect the necessary supplies of these nutrients from the poor herbage and hence, its growth is stunted. Again, when we speak of a highly productive dairy cow we really have in mind a type of animal that is capable of converting a large supply of nutrients into milk. Such an animal may be quite unsuited to a poor upland farm, because her milk production will tend to outrun her intake of nutrients and she will lose condition and powers of resisting disease. As a third example, the bacon type of pig is intended to produce a lean type of carcase with the least possible consumption of food. But if food is restricted in the early stages the development of lean tissue will be inhibited, and heavy feeding at a later stage will result in a large development of fat, so that the final product will not be what is desired. The expression of the animal's hereditary merits thus depends on the provision of appropriate living conditions.

When Bakewell,[1] the great pioneer breeder, started the improvement of British livestock about the middle of the eighteenth century, the farm animals of the country were of very mixed type and of a low average standard of merit. Certain local types were distinguished by peculiarities of size, prevailing colour, life., but there were no breeds in the modern sense. In all these local types there seems to have been a great amount of variability with respect to useful qualities. Probably there were many individuals with one or other good quality, and occasional specimens showed chance combinations of several valuable qualities; but ancestry was mixed and transmission apparently very irregular. The initial step in the improvement of each breed in turn, by one improver after another, has been to collect a herd or flock of the best individuals of the local type, and so to get possession of all the qualities that it was deemed advisable to perpetuate in the improved breed. If it happened that several of these characters could be obtained already combined in one individual, then the work of building up the desired combination would be rendered less intricate and tedious. But the main point is that the breeder could not expect, except as a rare piece of good fortune, to get any actual change in the individual unit characters with which he started-a favourable mutation. And hence, the real essential in the foundation of a herd was to get together as many good characters as possible, even if this involved the introduction of some bad. For bad characters could be "bred out," whereas there is no known method of producing good characters *de novo.* Apparently Bakewell, when he laid the foundations of his Leicester flock, succeeded in gathering together all the characters that were essential to the making of a good long-woolled sheep, mixed up as they were with a greater number of undesirable qualities. On the other hand, his work with the Longhorn cattle of the Midlands was destined to comparative

[1] Robert Bakewell of Dishley in Leicestershire, b. 1726 d. 1795, originator of the improved Leicester sheep and the improved Longhorn cattle, and breeder of cart-horses and pigs.

failure, apparently because elsewhere, in the valley of the Tees particularly, there was a supply of better raw material local type of cattle among which there lay scattered a set of more valuable unit characters.

Having started on a basis sufficiently wide to ensure the inclusion of all the characters desired, the breeder proceeds, by successive matings and selection of the progeny, to weed out undesirable characters, and to combine the good qualities first in an individual or two and finally in a true-breeding strain. This process, it is true, has never yet been carried to the point of complete or final success, yet so much progress has been made that it is important to survey the methods by which it has been achieved.

MASS AND PEDIGREE SELECTION

The simplest form of selection is what is termed mass or phe-notypic selection-the choice of breeding-stock on the simple basis of individual merit. The application of this process has given very different results according to the kind of characters to which it has been applied. Obviously the degree of success attained will depend on the accuracy with which the genetic make-up is reflected in the visible or measurable characters. If a certain visible character represents a certain definite and pure (homozygous) condition in the germ cells, then we are on perfectly safe ground. Thus, if we wished to obtain a race of single-combed fowls from a mixed population, a system of mass selection would give immediate, and completely successful results, because we know that a single-combed fowl can have in its germ cells nothing but the "single" factor in the homozygous condition. For various reasons, however, the visible characters of the animal may be unsatisfactory, or even quite misleading, as a guide to the qualities that it is capable of transmitting. Characters like egg-production in fowls or milk-production in cattle are determined in part only by heredity, being influenced to a very marked extent by the conditions of "nurture" in the widest sense. Even when, in the case of a milk cow, we calculate the allowances for the grosser effects of age, time of calving, feeding, etc., we are left with a multitude of varying environmental conditions whose total influence it is impossible to assess. We must, therefore, proceed on the assumption that milk-yield is in large measure a matter of accident or "pure chance," and is only in part determined by heredity.

Even when a certain desirable quality is known to be solely a matter of heredity, mass selection often fails to give the desired results. This happens if the character in question is dominant to a corresponding undesirable condition. For example, in cattle the red colour is recessive to most others, and is completely dominant to none [1]; hence reds breed true, and mass selection is immediately successful in establishing a true-breeding red strain With Angus cattle, on the other hand, the recognized black colour is completely dominant to red, *i.e.* there is no visible difference between the homozygous black, which breeds true, and the black x red heterozygote which throws reds. The consequence is that red calves occasionally turn up in most herds even though their ancestors may have been black for generations.

There is still another set of circumstances in which simple selection fails to accomplish anything. This happens when a particular quality is due to a hybrid condition of the germ-plasm Thus, the pink colour in *Mirabilis,* the blue of the Andalusian fowl, and also the roan and the blue-grey colours in cattle, are each due to a heterozygous condition, and it seems that one might breed from any of these types for countless generations and still be no nearer to a true-breeding strain.

Pedigree, when rationally used, forms a valuable aid to the breeder in his work of selection. Pedigree must never be regarded as an end in itself; individual merit must always be considered first-but the two taken in conjunction are undoubtedly a far better guide than either considered alone. The essential questions that a breeder should ask himself in attempting to evaluate a particular pedigree are: (1) What was the average merit of the animal's immediate ancestors? (2) What evidence is there that their good qualities were regularly transmitted from generation to generation? If we have a good individual, the progeny of good and true-breeding parents, it is probable that it will transmit its good qualities. The commonest error in the interpretation of pedigrees is that of attaching undue importance to far-back ancestors, and in particular to descent in the direct female line from particular "foundation females." It is quite certain that the average influence of any single animal in even the third or

[1] This statement does not apply to white markings. Some types of white pattern are recessive to solid colours.

fourth ancestral generation is very slight indeed, and it is doubtful whether in practice there is much to be gained by looking back beyond three generations.

GENOTYPIC SELECTION

The third and by much the most effective method is what biologists call *genotypic selection*-that is to say, selection on the basis of breeding ability, as determined by actual tests. The special value of this system arises from the fact that it enables the breeder to form a far truer estimate of the hereditary characters than he can do either on the basis of individual merit alone or on the dual basis of individuality and pedigree. We may illustrate the method by reverting to the previous illustration of the red Aberdeen-Angus. If it were a matter of prime importance to get rid of the red colour once and for all, the essential thing would be to breed from parents that were homozygous for black colour. If both parents were homozygous blacks, then the calves would all be homozygous blacks. But even if we made sure of one parent only, we should, for the time being, prevent the appearance of reds, and by continuous breeding from homozygous black on one side we should very quickly reduce the number of the black x red hybrids in the herd. In practice this is all that could be done, because it is impracticable to carry out breeding tests with females of the slower-breeding species; half the cow's breeding life might be over before one could get presumptive evidence of her germinal constitution, Genotypic selection would, therefore, be confined to the bull, and the procedure would be to mate him with, say, ten or a dozen red or white or red-roan cows. If he were a homozygous black, all his calves would be black or blue grey. If he left any red or red-roan calves one could definitely conclude that he was a black x red heterozygote, and therefore he would be discarded.

A similar procedure may be adopted with regard to any other character or group of characters. If the trial matings be made with carefully chosen females, and if the progeny be sufficiently numerous, a very complete picture can be obtained of the germinal constitution of the sire. Thus, the best test of the capabilities of a dairy bull is the average production of a number of his daughters compared with the average production of their dams. With meat-producing animals, too, the most satisfactory basis on which to judge a sire would be the economy of production and the quality of the meat produced by a number of his progeny. Such experimental tests are actually made with pigs by official testing stations in Canada, Denmark, and other countries; groups of four progeny of each sire are reared, fed, and finally converted into bacon under carefully controlled conditions, and the results are used in the final selection of stud animals from among the sires. In the United States, progeny testing by similar methods is being applied to bulls of the beef breeds. Such results are very valuable, for they constitute a test of the true commercial value of the animal under ordinary farm conditions.

FIXING TYPE

When there has been obtained an animal with something approaching the desired set of hereditary characters, the problem arises of maintaining or fixing this combination in a herd or flock or strain. It is true that no breeder has yet succeeded in producing a perfect animal, or even one possessing a full complement of all the best characters of its breed; yet success in the breeding of really fine individuals is comparatively common. It is far more difficult to build up a herd in which the good qualities are transmitted in any complete or regular fashion. Take milk-production, for example. Individual yields of 2000 and 3000 gals, and even more, have been obtained. The best herds, on the other hand, produce averages of between 800 and 1200, and this under a system of continuous culling-that is to say, the average is not that of the females bred but the average after a considerable proportion of the poorest producers has been discarded. In many countries the average production is 400 gals, or less. Admitting that milk-production is influenced by many circumstances other than heredity, one cannot avoid the conclusion that it is extremely difficult to get a fixed and true-breeding strain of heavy producers. It is obvious, too, that this end is of far greater importance than that of securing a few high individual records.

INBREEDING

It has been known since the time of Bakewell that in breeding the mating together of closely related animals-is a valuable aid in the fixing of type, *i.e.* in the production of strains in which the same characteristics are regularly reproduced. The discovery by Bakewell of the value of inbreeding seems to have been due to the accident of circumstances. He had new and original notions as to what constituted merit in a sheep or a steer, and after he had formed his flock and herd, and had gone some way with their improvement, he naturally found it increasingly difficult to obtain, from outside, sires that he considered good enough to introduce. He, therefore, turned to home-bred sires, and gradually, as his blood lines became intermingled, he resorted to more and more intensive inbreeding. On the whole his results were exceedingly good.

This effect of inbreeding, the fixing of type, is readily explicable on the Mendelian hypothesis. If we take a group of plants that are heterozygous for a particular character-say a hundred pink-flowered *Mirabilis-and* cause them to become self-fertilized, 50 per cent, of the progeny will be heterozygous, *i.e.* pink, while the other half will show one or other of the homozygous conditions, white or red. If this generation is selfed again the whites and reds breed true, while the pinks produce 50 per cent, of the hybrid (pink) and 50 per cent, of the homozygous (white and red) forms. Thus, if the various forms are allowed to multiply in the same ratio, the proportion of heterozygous or non-pure-breeding types decreases from generation to generation in the series 1/2, 1/4, 1/8, 1/16, and so on. After ten generations of self-fertilization the pure types would constitute 99.9 per cent, of the population. Similarly, if we start with a multiple hybrid-i.e. with a plant having a large number of characters in the heterozygous condition -and if we neglect the complication of linkage the *average* number of heterozygous characters will be halved each time that self-fertilization occurs. If, for example, we cross two varieties of wheat that differ in sixteen Mendelian characters, and allow the progeny to self-fertilize for a dozen generations, it may be calculated that over 99 per cent. of the progeny will have automatically reached the homozygous or true-breeding condition[1]. This is in accordance with the discovery, made-independently, that in species that are regularly self-fertilized the vast majority of individuals do actually breed true.

Self-fertilization is, of course, the most intense form of inbreeding that it is possible to conceive, but other forms differ from it not in kind but only in degree. Whenever inbreeding is practised, characters tend to

[1] In fact the occurrence of linkage slows down the rate of progress towards hmozygosity.

separate out in the homozygous condition, and the closer the degree of relationship the more rapid will tend to be the process of segregation into pure-breeding strains. Continuous out-crossing, on the other hand, tends to keep a maximum of characters in the heterozygous condition, with a correspondingly wide range of possible combinations and a high degree of variability.

It must be pointed out that absolute purity of type, in the Mendelian sense, has probably never been reached with farm animals. In order to isolate "pure lines" from an ordinary mixed stock we should have to breed very closely for a considerable number of generations. In the case of full-brother-and-sister matings it might take some twenty generations. Nothing approaching this amount of inbreeding has ever been attempted in practice,[1] and so far we can only speak of strains of farm animals as relatively pure. Even in the most carefully selected and most closely bred families there must remain a considerable number of heterozygous characters. That this should be so, in spite of the obvious advantages of pure-breeding types, is accounted for by another phenomenon connected with inbreeding.

Very generally, when attempts have been made to inbreed through several generations, the breeder has found that his stock deteriorated in some respect-showed a lack of size, or vigour, or both; susceptibility to disease; lessened fertility, or perhaps complete sterility. The largest controlled experiment with cattle was carried out in the United States with Friesians. One section of the herd was "outbred", each successive herd sire being a carefully selected proven bull which was as nearly as possible unrelated to the females. The other half was closely inbred, each successive sire being a son or brother of his predecessor.

The conception rate was three services per pregnancy in the inbred section as compared with 1.9 in the outbred. The birth weight of the calves fell from 81 or 82 lb. in the first generation to 63.9 lb. in the sixth inbred generation. Mortality of calves was much higher among the inbreds than the outbreds. Milk yield was maintained for the first three generations

---

[1] A very high degree of uniformity is obviously required in laborastory animals, such as rats and guinea-pigs, which are used for the testing of the strength of drugs and the vitamin potency of foods. Some strains of these animals have been so much inbred that they seem to have reached the condition of pure lines.

of inbreeding but afterwards declined. The cows of the sixth inbred generation gave 31 per cent, less milk and 38 per cent, less butter fat than the original foundation cows.

When we mate two animals or plants, belonging to different inbred strains, the cross-bred progeny often show markedly greater vigour than either parent. This phenomenon is called "hybrid vigour" or *heterosis.* The theories that have been advanced to account for the facts-the decline in vigour under inbreeding and the gain when two inbreds are crossed-fall into two groups.

On the one hand it is argued that selection for economic qualities in an outbred population will quickly eliminate unfavourable dominant factors, whereas an unfavourable recessive, unless it is carried by a large proportion of the stock, will escape notice and will persist. When inbreeding is practised, however, the homozygous recessives will segregate out and, if their number is considerable or their effects severe, the strain will "degenerate."

When we cross two unrelated (or only distantly related) inbred types the chances are that, with respect to the majority of factors, one or the other parent will contribute the favourable dominant, so that the cross-bred progeny will have few unfavourable recessive factors in the homozygous state. This theory appears to accord with most of the known facts. For instance, in the improvement of self-fertilized crop plants, the common procedure is to make a cross, to allow selling for a number of generations, and to build up a new strain consisting of the progeny of a single homozygous plant. Such completely homozygous families may be of high economic value. Again, with certain laboratory animals-rats, mice, and guinea-pigs-inbreeding has been carried to the stage of almost complete homozygosity, but so far this has not been achieved in the case of any type of farm livestock, the reasons for the difference being probably that it is possible to keep small animals in relatively large numbers, that the generations are shorter, and that the higher level of prolificacy gives greater scope for selection. But it is also to be noted that many of the inbred families of laboratory rats, etc., are smaller and less prolific than the out bred populations from which they have been derived.

Another hypothesis, that does not exclude the last, assumes that "over-dominance" (or super-dominance) affects the influence of certain genes-i.e. that a favourable dominant can be actually more effective in the heterozygous than in the homozygous state.

Whatever the theoretical explanation, it is clear that inbreeding does not lead to the creation of undesirable hereditary factors, but merely brings to light such as have been hidden. It does not lead, as was formerly supposed, to indefinite and progressive "running out"; deterioration comes to an end when the homozygous condition has been reached. Obviously, however, if the advantages of inbreeding, in the way of creating true breeding types, are to be secured without the disadvantage of loss of vigour, etc., inbreeding must be accompanied by rigorous selection, so that, in practice, the degree of inbreeding that should be attempted must depend on the scope for selection. This, in turn, depends on the reproductive rate of the particular animal with which the breeder is concerned; for instance, there is far more scope for selection in poultry than in dairy cattle.

What is called 'line breeding" is merely a less intensive form of inbreeding. The term is ordinarily applied in cases where the degree of kinship, between animals mated, is less close than that of full cousins. But statisticians, given an animal's full pedigree, can now place a numerical value on the degree of inbreeding. This expresses the percentage reduction, in the original number of heterozygous factors, that would be expected from the occurrence of particular ancestors at two or more places in the pedigree. It is rare to find, in ordinary pedigree herds, animals that are more than "10 per cent, inbred."

The most extensive experiments on the inbreeding of farm stock have been carried out with pigs under the auspices of the Bureau of Animal Industry of the United States Department of Agriculture.

Work on farm animals was preceded by pilot experiments carried out by Sewell Wright with guinea-pigs, which began in 1906. Of thirty-five original families, twenty-three survived. In all but one of these every mating was between full sibs (brother and sister), and in the one remaining family was between parent and offspring. In all cases there was selection of the best individuals from each litter. An outbred population was maintained as a control.

The inbred families, without exception,showed a decline in all the characteristics measured-adult size, growth rate, prolificacy, etc., the most marked deterioration being in litter size. As would be expected, different families became differentiated in regard to coat colour and other characters. In the later years, crosses were made between representatives of different families, and all the cross-breds were markedly superior in every respect to either of the parent stocks.

This work suggested that the technique that has been successfully applied in the production of hybrid maize might also be applied to livestock.

The pig-breeding investigation has been conducted on a very large scale. No less than 112 inbred lines were started, at ten different State Experiment Stations, in the thirties, and the work has been coordinated centrally since 1937. Up till 1953 some 14,000 experimental litters had been bred, and the number of experimental animals, in the herds, had reached about 1500. The main criteria in selection have been prolificacy of the sows, the viability of the piglets, growth rate, efficiency in food conversion, and carcass quality. Most of the foundation stock for the inbred lines was pure-bred, but some families were founded on first-cross or second-cross material.

The heritability of the desired qualities was found to be fairly low-averaging, for example, about 16 per cent, for litter size (numbers born).

Hereditary defects became prevalent in certain lines-e.g haemophilia (bleeding) and a high incidence of scrotal hernia and the lines affected were discarded. Other families had to be abandoned owing to an extreme fall in prolificacy.

The intensity of inbreeding varied as between the various lines, most of the surviving families showing figures between 25 and 50 per cent.; but one line, produced by eight generations of brother x sister matings, had reached 80 per cent, inbreeding.

The average decline in prolificacy has been at the rate of about 0.3 pigs per litter for each 10 per cent, increase in inbreeding, but the variation as between families has been -high.

None of the inbred lines, as such, is useful for commercial production, but performance is better in two-line and three line crosses. Moreover, boars of certain inbred lines have proved very valuable for mating with sows of ordinary breeding, especially in cases where the boar and sow were of different breeds.

In general, the gain on crossing, in regard to useful qualities, has been greatest where the genetic differences between sire and dam were at their maximum The following figures show the average results obtained from crosses between an inbred boar and non-inbred sows of a different breed, as compared with those obtained by using a non-inbred boar of the same breed as the sows (i.e. ordinary pedigree breeding).

	Top Cross with Inbred Boar	**Non-top Cross**
Number of piglings farrowed	9.85	8.69
Number of weaned farrowed	7.64	6.56
Weight of litter at weaning	203.6	166.4

A special Association has been formed for the registration of animals belonging to the inbred strains. The most numerous of the new breeds is Minnesota No. l, of which in 1952 there were some ninety herds comprising about 15,000 registered animals.

The tentative conclusion to be drawn from the American experience would seem to be that there are definite advantages to be gained by the systematic production of inbred families and the subsequent use of the inbred males for crossing, but that an immense expenditure, over a long period, will be required for the full exploitation of the system. The production of inbred stocks of Large White and Wessex pigs is in progress at the Animal Breeding Research Organisation, Edinburgh.

GRADING AND CROSSING

Where a breeder starts, as is often inevitable, with a collection of females of average or inferior quality he should normally endeavour to "grade up" his flock or herd by the continued use of pedigree sires. Obviously, before he begins operations, he should convince himself that one or other of the pure breeds will provide the type of animal he desires, and having once decided on a particular breed he should not lightly discard it for another. Certain of the Breed Societies encourage such grading by admitting to their registers "grade" animals with a certain proportion of non-pedigree ancestry. So long as the breeder is in the grading up stage, *i.e.* so long as he is able to obtain from outside sources sires that are reasonably certain to improve the standard of merit in his herd-there is no particular advantage to be gained by close inbreeding; on the other hand, if he happens to secure a particularly good and impressive sire he should not hesitate to concentrate his "blood," and the experiment of mating him to his own progeny will often be worth trying. A breeder who is grading-up may study pedigrees with just as much advantage as the pedigree breeder himself, for if he is content to buy any registered sire on his appearance alone he may get an animal of nixed or inferior ancestry that may bring him serious loss.

Cross-breeding, so long as it is carried out systematically, has, for the production of "commercial" animals, certain definite advantages over pure-breeding. Cross-bred animals commonly show the phenomenon of *heterosis* or hybrid vigour that is to say, are more prolific and more resistant to disease and grow more rapidly than pure-breds. Theoretically, of course, it is possible to collect a complete set of desirable dominant factors in a single homozygous strain, but many generations must elapse before this can be accomplished with farm animals. Meantime, such valuable dominants as have been fixed in two breeds can be combined by crossing. Heterosis is usually most marked in the first cross; hence cross-breeding should not be carried beyond this stage or, at the most beyond that of mating a first-cross female with a pure-bred male. In the latter case the disadvantage of irregularity in the second-cross generation may be outbalanced by the advantages of hardiness and prolificness in the cross-bred dams.

In practice cross-breeding offers the greatest advantages in highly prolific species like pigs and poultry, because in these cases a relatively small proportion of pure matings will suffice to produce the necessary supply of selected pure-bred females. In the less prolific species a large proportion of the females born must be kept for breeding, and widespread crossing involves the retention either of cross-bred females or of inferior pure-breds. In the case of dairy cattle, crossing is very liable to degenerate into "mongrelizing," because the cow must breed in order that she may milk. Normally, grading is the better system to pursue in dairy herds.

It is a common misconception that inferior pure-bred sires are good enough for crossing. It is true, of course, that there is a limit to the sum of money that can be profitably invested in a crossing sire, and it is also true that animals which are faulty to the eye of the pedigree breeder may be excellent from the point of view of the commercial breeder. But it is only by careful selection of his sires, and by careful study of his results,' that the cross-breeding farmer can hope to repeat his successes and avoid recurrence of his failures.

CHAPTER II

THE PRINCIPLES OF NUTRITION

A CONVENIENT approach to the study of the nutrition of farm animals is to consider the various chemical compounds which are contained in feeding stuffs and the uses to which they are put in the animal body. The many hundreds of chemical compounds in feeding stuffs are classified by chemists into four major groups. These are: (1) mineral or ash constituents; (2) the nitrogen containing constituents; (3) the fatty and waxy materials; and (4) the carbohydrates. This classification is a very general one, and each major class contains a multitude of compounds; indeed, the chemical constitution of some is not exactly known. One very important component of feeding stuffs has, however, been left out of the above classification, namely water. The first step a chemist takes in the analysis of a feeding stuff is to determine its moisture-content by heating it at the boiling-point of water. He then determines the amount of the other major classes of material in the dry matter. Even the so-called dry feeds contain water. Freshly purchased concentrate cubes contain the least, namely about 10 per cent.; hays and straws about 10 to 15 per cent; grass and green forages between 65 and 85 per cent. The highest water-contents are found in turnips, and swedes, which often contain over 90 per cent. A high water-content, of course, increases the cost of handling and transporting a crop, but at the same time succulent foods, such as green crops, roots, and silage appear to have a special value which cannot be simply accounted for. A good supply of water for all farm stock is essential, because water is constantly being used to excrete, in solution in the urine and faeces, unwanted waste products arising from the breakdown of food materials in the body. A further loss of water occurs through the skin and in the breath of the animal; in fact a highly fed dairy cow will lose as much as 4 gals of water per day by evaporation. In the breakdown of food some water is formed in the body, but this amount goes only a small way to meet these losses. Water requirements of stock are met not only by drinking water but also by the water contained in their feed. In the spring, for instance, heifers grazing on lush pasture receive sufficient water from the herbage they consume, and do not go to the drinking trough. The same is true of sheep which on fresh pasture drink no water, but on old pasture, in dry weather, may drink up to half a gallon per head per day. The following table summarizes the normal water intakes by farm animals. With the exception of young calves, which never know when they have had enough, and fattening pigs, where a convenient rule is 3 lb. of water to 1 lb. of meal, farm animals are best left to judge their own requirements, and unlimited water should be made available.

Water Requirements on Dry Rations

Class of Stock	**Weight (lb.)**	**Water required per Day (lb.)**
Weaning calves	150	10 to 15
Dry cows	1000	70 to 80
Milking cows *	1000	Up to 180
Sheep	150	Up to 10
Weaning pigs	30	4
Baconers	200	8
In-pig sows	400	10 to 12
Suckling sows	300	40 to 50

* The milking cow requires it to 2 gals, of water for every gallon of milk she produces besides that which she would need if she were dry.

Minerals : The ash constituents of feeding stuffs are determined by incinerating the dry material at a dull red heat in a furnace. The unburnt residue consists of silica and the sulphates, chlorides, phosphates and carbonates of calcium, magnesium, potassium, sodium, iron, and a large number of other elements. Some of these others, like copper and cobalt, are of considerable nutritional significance. Others, like boron, aluminium, and nickel, while known to be present in minute amounts in animal tissues, do not appear to be essential. The amount of ash in feeding stuffs varies considerably. Some feeds, such as flaked maize, contain very little indeed-less than 1 per cent.-while feeds, such as fish meal contain more than 20 per cent. Most common feeds contain from 2 to 8 per cent, of ash, calculated on a dry-matter basis.

All animals require minerals, and some of these must be given in organic form. Thus, the animal must receive at least part of its sulphur requirements in the form of the sulphur-containing amino acid methionine. Some mineral elements are toxic when given in large amounts. Thus, common salt may easily poison poultry and pigs if inadvertently they are given doses too large in relation to their water supply. Again, cattle which have access to pastures which have received heavy dressings of phosphate fertilizers and have not been washed by rain may suffer from scours for a day or two. Some minerals are toxic even when only very small amounts are supplied. Molybdenum, which is present in the herbage of certain pastures in Somerset-the so-called "teart pastures," causes severe scouring in cattle. Luckily, the trouble may be prevented by giving cattle a special cake containing copper. Selenium is responsible for a disease of cattle in certain parts of Ireland and the United States. In this case, unfortunately, there is no antidote to the disease and, since selenium is present in the soil of the areas and in all crops grown thereon, it is impossible to carry on any system of livestock husbandry without running the risk of serious loss. Poisoning by fluorine, which may contaminate pastures through flue gases and industrial smoke from certain modern industries, produces widespread abnormalities of the bones and teeth of cattle and has caused many deaths in several areas of Britain. Of all toxic elements, lead is probably the most important in practice. Every year in Britain many hundreds of calves die as the result of chewing the lead-painted doors of their boxes.

Though certain minerals can cause ill-health or even death, much more harm to the productivity of livestock results from a shortage, in their daily rations, of essential minerals. Of these, calcium and phosphorus are of supreme importance. About 75 per cent, of the total ash of the body of farm animals consists of these two elements. Together they make up 99 per cent, of the ash of the bony skeleton, and phosphorus is also contained, in appreciable amounts, in the soft tissues. It is, thus, the rapidly growing young animal, which is laying down both soft and skeletal tissues, that has the greatest need for these elements. But milking animals also have a great need, since a gallon of milk contains about 1/5 oz. calcium and 1/6 oz. phosphorus. These amounts must, therefore, be supplied in the feed. But during early lactation the milking cow draws part of the requirements from her own bones and, later in lactation, rebuilds her depleted bones in preparation for her next lactation. A shortage of calcium and phosphorus leads to rickets and other bone abnormalities. The soils of some large areas, notably in South Africa, are by nature grossly deficient in phosphorus, and the natural herbage is so deficient that cattle grazing it develop the disease “aphosphorosis,” which is characterized by a stunting of growth and a depraved appetite. Mild forms of this deficiency disease have been seen in Britain, and it has recently been suggested that shortages of phosphorus, or an imbalance of calcium and phosphorus, may lead to infertility in dairy cattle.

In meeting the requirements of animals it must be remembered that most concentrates are low in calcium, and that some starchy feeds are very poor in phosphorus as well. The usual homegrown mixtures for milk production are generally deficient in calcium, phosphorus, and common salt, so that a proprietary mineral mixture or (more cheaply) a mixture of two parts common salt, one part ground chalk, and one part sterilized bone-flour should be added to the concentrate ration at the rate of 3 lb. minerals per cwt. Attention to the calcium and phosphorus supplies of pigs is also essential when these are kept and managed under intensive conditions. If the ration is reasonably good the addition of ½ per cent, ground chalk and 1 per cent common salt will suffice.

Trace Minerals : Animals that are living under natural conditions, *e.g.* on normal good pasture, are almost certain to receive sufficient mineral nutrients; but if the soil is seriously deficient in a particular mineral, deficiency disease will appear in the grazing animals. Disease will also occur in animals maintained on the conserved forage, in the absence of a supplementary source of minerals. One example is “pining” or “moor sickness” in sheep, which occurs on many hill grazings in Scotland and Wales and in certain parts of England also. Sheep confined to these hills or moors lose appetite and become anremic and weak The disease is due to the absence from the soil, and hence from the herbage, of the minute amount of cobalt which is necessary to the well-being of ruminants. The amount needed by the sheep is extremely small-about 1 part in 10 million of its food. Copper deficiency also occurs on some soils and results in a pining condition, or in scouring and unthriftiness, in cattle. In the pregnant ewe shortage of available copper results in her giving birth to lambs with “swayback.” Such lambs may be unable to stand, or can stand and walk only unsteadily, with a characteristic swaying of the hind quarters. In certain parts of Britain (notably in Derbyshire) these conditions occur even where the copper-content of the soils appears to be normal, and it is thought that the presence of other minerals in the herbage may make the copper unavailable. A further trace element deficiency which occurs in farm practice is that of iron and copper together. These elements are needed for the manufacture of haemoglobin, the red pigment in blood cells. Sows milk is grossly deficient in these elements, so that sucking pigs that are reared indoors often become aneemic and may die. Dosage with a solution of iron and copper both prevents and cures the disease. Piglets running out of doors can obtain iron by rooting in the soil and are, therefore, rarely affected. Deficiencies of manganese and iodine have also been known, but only rarely, to occur in farm livestock in this country.

In general, shortage of trace elements is rather rare, occurring only over a very small range of soil types or under highly artificial systems of intensive rearing. Provided that sufficient calcium, phosphorus, and common salt are present in mixed rations and that a fair variety of feeding stuffs is used throughout the year, there is seldom any danger of deficiencies of the other mineral elements.

The Nitrogen-containing Constituents : Returning to our broad classification of the constituents of feeding stuffs, we come to those which contain nitrogen. These include a vast number of different substances, some being of fairly simple constitution, while others are of very great complexity. The most important members of this latter class are the proteins, such as the casein of milk, the albumen of egg-white, and gluten, one of the main proteins of wheat grain. Sometimes this whole class of nitrogen containing nutrients is referred to as *crude protein,* and a distinction is made between the actual or *pure protein* and the remainder, sometimes termed *"amides"* but more correctly, *non-protein nitrogen-containing substances. True proteins* contain some 15 to 17 per cent, of nitrogen and are of exceedingly complex composition. When broken down, an individual protein may yield upto twenty or more relatively simple compounds called *amino-acids.* Of these twenty a proportion can be manufactured in the body from quite simple forms of organically combined nitrogen, but ten of them, the so-called *essential amino-acids,* have to be provided in the non-ruminant animal's diet in order that it may be able to build up its own body proteins, those of eggs and of milk and the keratin proteins of hair. The non-proteins of feeding stuffs include firstly amino-acids and amides, and (in smaller amounts) alkaloids, lecithins, betaines, and many other simple crystalline materials. While some of these simple compounds can be used or broken down by the animal, it is only the amino-acids, whether derived from the proteins or from the non-protein fraction, that are essential.

The proportion of the crude protein which is in the form of pure proteins varies greatly from one feeding stuff to another. In ripe seeds it is commonly more than 90 per cent, while in mangolds it is only about 10 per cent, the remainder, in this case, including some nitrate. In rapidly growing green crops, in which the manufacture of protein is proceeding, as much as 50 per cent, of the nitrogen is represented by non-proteins, such as the amides asparagine and glutamine. The total content of crude protein also varies widely; for instance, cereal straws contain only 1 or 2 per cent, beans and peas about 25 per cent, and some oil cakes contain more than 40 per cent.

In the digestion of protein-containing feed by simple-stomached animals, such as pigs, the enzymes of the digestive tract break down the proteins to amino-acids. These are absorbed into the blood stream and are later re-assembled in the tissues to produce the specialized proteins of muscles, liver, hair, milk, and so on. The proteins vary in composition, yielding various proportions of the common amino-acids. Some proteins fail to yield particular amino-acids; for example, gelatine, from bones, yields none of the amino-acid tryptophan, while gliadin, from wheat grain, yields no lysine and very little tryptophan. Information is scanty but it is known, for example, that the proteins of the pulses tend to be low in lysine, and that the best of all, from the point of view of amino-acid make-up, is the mixed proteins of whole egg. If the pig be given an otherwise adequate ration, with gelatine as the only protein, it fails to gain in weight because of the lack of the essential tryptophan. Again, if given a diet with wheat proteins as the sole source of protein, its growth rate will be low because of a shortage of lysine. The pig, in re-assembling the amino-acids derived from its food, is limited by the supply of each individual one. Those which it cannot use are broken down in its liver, the nitrogen being converted first to ammonia and then to urea, which is excreted in the urine. The carbon containing residue can, however, be used as a source of energy and, in part, stored as body fat.

Food proteins, thus, differ widely in their ability to produce body gains in the pig and other simple-stomached animals. The degree of efficiency of utilization of a protein is usually expressed as its *biological value.* Proteins with a biological value of 100, such as those of whole egg, are such as can be utilised by the animal without wastage. A protein, such as gelatine, when fed by itself, can hardly be used at all, and has a biological value of about 25. The proteins of cereals are about 70 per cent, as good as whole-egg protein, while those of skimmed milk are nearly as good, having a biological value of over 90. In general, animal and fish proteins. such as those of blood meal, meat meal, and while-fish meal, are much superior to plant proteins as sources of essential amino-acids. It is broadly true that a mixture is better than any single protein, since a deficiency of a particular amino acid in one protein is often balanced by an excess in another. For example, better growth is obtained if an all-cereal ration, which is 'deficient in lysine, is supplemented with gelatine, which yields a large amount of lysine. This is true despite the fact that gelatine provides none of the essential tryptophan. The biological value of the protein is unlikely to be a limiting factor in the growth of pigs and poultry if mixed rations are employed, particularly if, as is usual, some fish meal or other animal protein is included in their rations.

With pigs and poultry it is quite certain that the amino-acid composition of the food protein is important in relation to growth. By contrast, the importance of biological value to ruminants is much less.

In the ruminant all food first enters the capacious paunch, which can be described as a fermentation vat. Here vast numbers of micro-organisms break down the proteins into their constituent amino-acids and indeed often beyond this point. Simultaneously they rebuild the fragments into their own body proteins. The partly fermented food mass, containing the bacteria, then moves down the digestive tract where the proteolytic enzymes hydrolyze not only the food proteins which have escaped ruminal fermentation, but also a proportion of the bacterial proteins as well. Thus, the cow and the sheep assimilate not only the mixture of amino-acids supplied in their food, but also the complex mixture which is the product of bacterial action in degrading and synthesizing amino-acids from those provided by, or derived from, the non-protein nitrogenous compounds of the feed. The ruminant is indeed able to utilize quite simple nitrogen compounds, such as urea and ammonium salts, as a partial substitute for protein, the explanation being that the bacteria can use these to manufacture amino-acids. It is, thus, clear that, for ruminants, the biological values of proteins have only a limited significance; first-class proteins tend to be degraded and poor proteins to be upgraded, so that only small differences in growth, protein storage, or protein synthesis by the animal result from widely different forms of nitrogen compounds. There is, however, strong evidence to the effect that the rumen bacteria are not capable of providing the ruminant with all the amino-acids it needs. In particular, they are not very efficient in the manufacture of the sulphur containing amino-acid methionine, and it is improbable that they can provide certain particular amino-acids in amounts adequate to provide for a high level of milk production. Some early experiments certainly suggest that a shortage of lysine in the ration can cause a fall of milk yield.

The facts set out so far to explain the observation that dairy cows show very little response, in milk yield, to major changes in the make-up of their food protein, and suggest that, for ruminants, the non-protein nitrogen fraction of feeds has a value but little short of that of true protein. In British feeding standards some cognizance is taken of these facts by ascribing to the digested non-proteins half the value of pure proteins; the sum of the digested pure protein and half the digested non-proteins is called the "protein equivalent." A sample of swedes, for instance, might contain 1.2 per cent, of digestible crude protein, consisting of 0.2 per cent, of digestible true protein and 1 per cent, of digestible non-proteins. The protein equivalent of swedes would, thus, be $0.2+\frac{1.0}{2.0}0.7$ per cent. It would seem, however, that the protein value of foods for ruminants would be better measured by the total figure for digestible crude protein, without correction.

The Fats : The next large group of food constituents are the fats. These, as determined by the standard method of analysis, include not only the true fats or glycerides, but resins, organic acids, essential oils, sterols, and plant pigments, such as chlorophyll and the carotenoids. All these constituents are extracted from the dry, finely-ground feeding stuffs by prolonged treatment with hot ether, and hence, they are more correctly known as "ether extractives." In some foods, such as linseed, the glycerides account for the larger part of the ether extractives, while in others-notably the straws and hays-the non-glycerides are the more abundant. In general, the glycerides have a much greater feeding value than the non-glycerides, but the latter fraction includes certain very important compounds, notably vitamins A, D, and E, which, as we shall see later, are of paramount importance in nutrition.

The Carbohydrates : The carbohydrates of feeding stuffs are generally classified as follows :

1. *Monosaccharides,* or simple sugars, such as grape sugar and fruit sugar, which are hexoses with the general formula $C_6H_{12}O_6$ and the pentoses with the general formula $C_6H_{10}O_5$
2. *Disaccharides,* which include cane or beet sugar, milk sugar, and malt sugar, and have a general formula $C_{12}H_{22}O_{11}$
3. *Polysaccharides,* which make a very large group of compounds, all of which have a high molecular weight and may be regarded as complex condensations of the simple sugars. They include starches, which are the reserve materials found in many plants, and dextrins, the simpler compounds which arise during die build-up or breakdown of starch. In digestion by animals with, simple stomachs (pig and fowl) nearly 100 per cent, of the starches, dextrins, and disaccharides are hydrolyzed to simple sugars by the action of enzymes secreted into the digestive tract. These simple sugars are then absorbed into the blood stream. The other important group of polysaccharides are those which are generally known as "cell wall constituents" or, more loosely, as "fibre." They include a very large number of compounds pectins, cellulose; lignin, hemicelluloses, pentosans, and polyuronides. Of these, the most important are cellulose and lignin. Cellulose is a major component of cell walls and is a polymerized glucose, yielding glucose on hydrolyzation. However, none of the enzymes produced by the wall of the digestive tract can carry out this hydrolysis, and thus, simple-stomached animals cannot digest cellulose. But the immense number of bacteria which are present in the rumen of cattle are able to break down cellulose and, in doing so, supply their host with a source of food it could not otherwise obtain. A little fermentation of the same type does take place in the caecum of simple-stomached animals, accounting for the small amount of cellulose which is apparently digested by these species. The product of the fermentation in the ruminant is not, however, glucose. When cellulose is fermented by the cow, the gas methane is formed and simple fatty acids, notably

acetic, are produced as well. Some of the cellulose is converted into bacterial starch in the bodies of the organisms, and some of the energy locked up in the cellulose is liberated as heat. Hence, the ruminant does not obtain the whole of the energy of the cellulose which the bacteria break down. The gas methane is of no value, the heat produced being often (though not always) a burden rather than an advantage, while some of the bacterial starch may probably escape digestion in the lower part of the digestive tract. The ruminant can, however, utilize the lower fatty acids, and experiments have clearly shown that these can be directly incorporated into milk fat. The bacteria can also attack starch and the simple sugars contained in the diet.

The other important cell-wall constituent, lignin, is neither attacked by bacteria at all, nor is there any digestive juice which can break it down. Lignin is, thus, completely inert. This is of very great importance in the feeding of livestock, since as plants get older lignin is laid down on top of the cellulose layers in order to give rigidity to the stem, leaf veins, and seed coat. This process of lignifications makes the plant less nutritious because the cellulose and the cell constituents are protected by the coating of lignin from the action of the bacteria. Lignification accounts for the decline in the digestibility of hay, when cutting is delayed.

The usual way of measuring the amount of woody fibre, *i.e.* lignified cellulose, is to boil the ground material first in dilute acid and then in dilute alkali. The insoluble residue is called "crude fibre." This represents only a rough approximation to the lignin and cellulose content, because lignin is partially soluble in the alkali and cellulose partially soluble in the acid. Thus, "crude fibre" from a late-cut hay or a straw is very different, in terms of nutritive value, from the "crude fibre" of young spring grass. The latter is extremely well-digested by cattle and sheep but the "crude fibre" of straws is not. The chemical method of estimating the other carbohydrates is even more unsatisfactory. They are calculated by difference, *i.e.* the sum of the ash, nitrogen-containing materials, ether extractives, and crude fibre is deducted from the total dry matter. The difference is called "carbohydrates" or, more correctly, "nitrogen-free extractives." No entirely satisfactory scheme of analysis for true carbohydrates has yet been devised. The best available measure of the value of the total carbohydrates of a feed is obtained by feeding experiments in which the digestibility of the crude fibre and nitrogen-free extract is determined.

The Vitamins : In 1906, the late Sir Frederick Gowland Hopkins stated that "no animal can live on a mixture of pure protein, fat, and carbohydrate even when the necessary inorganic material is supplied." From his own experiments and others, Hopkins concluded that some other substance or substances were required. These were first called "accessory food factors" and, later, vitamins. Today there is an immense amount of knowledge not only about their chemical structure but also their physiological effects both on laboratory rats and guinea-pigs and on farm animals and man. The vitamins are classified as either fat-soluble or water-soluble. The fat-soluble group includes vitamins A, D, E, and K, and the water-soluble group vitamin C and the numerous members of the vitamin B complex.

Fat-soluble Vitamins : *Vitamin A* is a colourless substance of known chemical constitution which is found in the liver oils of fish and animals. It is formed in the animal body largely from the orange-yellow beta carotene, one of the several yellow pigments called carotenoids associated in the plant with the green colouring material chlorophyll-carotene is indeed not the only precursor of vitamin A. Yellow maize grain contains crypto-xanthine which is also a precursor of vitamin A. But -carotene is the most common. All animals have the power of converting carotene into vitamin A, but some do so more efficiently than others. Thus, the Guernsey cow produces a yellow milk fat because she passes into her milk a considerable portion of the carotene she ingests. The Ayrshire, on the other hand, given exactly the same ration, converts more carotene into vitamin A before secreting it. The goat secretes no carotene in her milk. Despite the difference in colour the values of these milk in preventing disease due to vitamin A deficiency can be the same. No animal can manufacture vitamin A in its body except from the precursors, and the vitamin A potency of eggs and milk is directly proportionate to the amount of carotene which the hen or the cow obtains from its food.

As would be expected, carotene is largely found in green, leafy materials. Seeds and common farm concentrates contain very little, even yellow maize being a rather poor source. Carrot roots are very rich in carotene -but mangolds, swedes, turnips, sugar-beet pulp, and potatoes supply practically none. Some fish meals contain a little pre-formed vitamin

A, but the small amount is sometimes lost during processing. The green leaf is the main and the richest source, and the carotene-content falls markedly as the plant matures. Carotene is not very stable and is rapidly destroyed in the presence of oxygen. Carotene in cut grass is, thus, largely destroyed in the common process of hay-making, and it is only if the hay retains a good green colour that it can be regarded as a highly potent source. The table shows the original carotene-content of a crop of grass and indicates the losses which occur when it is conserved in different ways.

Material	Carotene-content of Material on a Dry Basis, mg. per 100 g. (Parts per 10,000)
Fresh grass	30
Hay made with little weathering	8
Hay, badly weathered	4
Ensiled	20
Artificially dried	26
Hay, badly weathered, after storage for two months in bales	1

Artificial drying is clearly the best method of conservation from the point of view of conserving carotene. Bales of dried grass, however, gradually lose their carotene-content through oxidation, this being particularly noticeable on the outside layers. Under most circumstances, dried grass of high quality is the best source of the vitamin. Cod-liver oil, which supplies pre-formed carotene, has been much used in the past, but when mixed with feeding stuffs its vitamin A value falls off very rapidly, complete destruction occurring withirt about three weeks. Moreover, under some conditions cod-liver oil is now known to have bad effects on animals. It is, therefore, generally advisable to rely on young green material for meeting vitamin A requirements.

To ensure an adequate supply, it is necessary for the farmer to maintain a supply of fresh, leafy green stuff, such as kale, clover, lucerne, or grass throughout the year, and to have a reserve of high-quality silage or dried grass. Foods that are rich in the vitamin need be given only in small amounts. Three ounces of really young leafy grass will meet the daily requirements of a growing pig and 2 to 3 lb. those of a milking cow.

During the summer and autumn, when greenstuff is plentiful, animals are able to store both vitamin A and carotene in their livers and are able to fall back on these stores during periods when their diet is deficient in the vitamin e.g. is composed largely of mangolds, straw, corn, and cake. Thus, if a calf, which has been running with its dam on pasture, is brought indoors in autumn and is given a ration completely free of the vitamin, it will have sufficient reserve to last for at least five months. Similarly, a pig, given a single large dose of vitamin A concentrate at weaning, will store enough in its liver to last it until it reaches bacon weight. In milking animals, however, there is a steady drain on the reserves, and these are much more rapidly exhausted. This does not mean that the cow will show symptoms of deficiency; in fact, it is necessary to feed a cow on carotene-low rations for well over a year before failure of reproduction and the general breakdown occurs. She may, however, give birth to a weak or dead or blind calf before she has begun herself to show symptoms of deficiency.

Young animals, in general, are born with very small reserves of vitamin A in their livers. Moreover, they appear to need more of the vitamin, in relation to their body weight, than do older ones; but they obtain a large amount of the vitamin from the colostrum of their dams, which is a much richer source than full-lactation milk. The vitamin A potency of colostrum-due both to preformed vitamin A and to carotene-is influenced by the carotene-content of the ration which the animal receives during pregnancy. Thus, after a long winter in which rations low in carotene-content have been fed, the colostrum will have a low vitamin A value. It will, thus, be clear that the proper feeding of cows during pregnancy is of great importance from the point of view of the health and viability of the calf.

Vitamin A deficiency in young cattle is not uncommon in this country. The main signs are failure to thrive, night blindness, and, later, total and incurable blindness due to damage to the optic nerves. In pigs, vitamin A deficiency leads to paralysis of the Mud quarters and a staggering gait. In chicks, growth is retarded, the birds become weak and emaciated, and the membranes of the mouth become thickened and pustular. The eyes, too, are often seriously affected. In all animals secondary bacterial infections occur where the deficiency is prolonged, and these may lead to the death of the animal. In this connection there is a great deal of evidence to suggest that many calves, in late winter, have subnormal reserves of vitamin A, and it is possible that the high incidence of scours, pneumonia, and other transmissible calf diseases at this season may be accounted for by shortage of vitamin A.

Vitamin D was first identified as a substance, present in cod-liver oil, which was essential to the prevention of rickets, vitamin D is, in fact, often called the anti-rachitic factor. It is now known that a number of plant and animal sterols, when exposed to the ultraviolet rays of sunlight, become potent as vitamin D. In all, there are at least ten different forms of vitamin D, the most important being vitamin D_2 (the form found in plant products, such as sun-cured hay and irradiated yeast) and vitamin Da, which is found in animal products, notably in fish oils, and indeed is formed in the animal body itself during exposure to ultra-violet light. The distinction between vitamins D2 and Da is important because, while both have the same antirachitic value for the laboratory rat and other species, vitamin Da is far more effective than vitamin D2 for poultry. Fish oils and irradiated animal sterols are, thus, much better sources for poultry than are irradiated plant materials.

Vitamin D itself is very sparsely distributed in animal feeding stuffs, but the sterols are fairly widely distributed. Fresh green' foods contain very little vitamin D. For example, in rapidly growing pasture grass it is only small dead areas of leaf, which have been irradiated by the sun, which contain any. For the same reason artificially dried grass, not having been exposed to the sun, has a very low vitamin D content. On the other hand, sunshine and "sky-shine" irradiate the skins of animals running out of doors, and hence, produce in them a supply of the vitamin. The problem of vitamin D deficiency, thus, concerns animals which are kept indoors and do not receive sun-cured foods. In this connection it should be Loted that ordinary window glass does not allow the short wave-lengths of ultra-violet light to pass through it; and hence, though a building may be well lit, the glass-filtered light will not prevent rickets. Glass substitutes are, however, available which, provided they are kept clean, admit the ultraviolet light.

Unlike vitamin A, vitamin D is not very easily oxidized; in fact, vitamin D was discovered by bubbling oxygen through cod-liver oil, thus destroying vitamin A and leaving the vitamin D unimpaired. The vitamin D activity of hays neither therefore, decline during storage, nor does the activity of vitamin D concentrates deteriorate when they are nixed with other feeds.

Rickets, the disease due to vitamin D deficiency, is characterized by a failure of the animal to lay down calcium and phosphorus in its bones. The cartilaginous material is not calcified, the bones become rubbery, so that bent and malformed limbs result. Rickets can, however, occur as the result of a deficiency of calcium or of phosphorus in the diet, and it is necessary to the understanding of bone growth to consider calcium, phosphorus, and vitamin D together. When the ratio of calcium to phosphorus is abnormal, then vitamin D will often permit normal growth, whereas if the ratio is optimal it may be very difficult to demonstrate any requirement for vitamin D. Vitamin D is not a substitute for calcium and phosphorus, but it enables growth to take place under dietary conditions which would otherwise soon result in unsatisfactory bone formation. Hens given rations low in vitamin D produce fewer eggs of lowered hatchability, and the few eggs that are laid have very thin shells. In chicks, rickets is very common and can be recognized by the ungainly attitudes that affected birds adopt to balance their bodies, and by their lame, stiff-legged gait. The breastbone is often bent and the beak becomes soft, rubbery, and easily deformed.

Vitamin E, when first discovered, was found to be a dietary ingredient essential for the reproduction of the rat. Without it male rats developed permanent sterility, and in pregnant females the embryos degenerated and were re-absorbed. Experiments and practical tests have shown that with cattle, sheep, and goats vitamin E is apparently not concerned with reproduction. Pigs may require it for normal reproduction, and it is known that adult birds, if deprived of it, produce eggs which have a low hatchability.

Though not concerned with reproduction, the vitamin is nevertheless of importance in farm animal nutrition. Under certain conditions associated with rations low in vitamin E content, cattle and sheep develop a muscular disease called muscular dystrophy or "stiff-lamb disease," while poultry develop severe nervous and circulatory disturbances. All these diseases have been found to occur on farms in Britain where inadequate rations were given. One important aspect of the disease in calves is that it follows dosing with cod-liver oil. Though cod-liver oil supplies vitamins A and D it contains unsaturated fatty acids which increase the animal's vitamin E requirements as much as fifty-fold. Small amounts of most proprietary concentrates of vitamins A and D do not have this effect. Vitamin E is found in large amounts in green foods and, as with vitamin A, the chief danger of deficiency is in young animals born in the later winter to dams which have received poor rations throughout a considerable time.

Vitamin K is concerned with the maintenance of the clotting time of the blood. An animal deficient in this factor may bleed to death from a minor wound. The vitamin was discovered in experiments with chicks. In parts of North America deficiency of vitamin K occurs when cattle eat mouldy *Lespedeza* hay. The symptom in cattle is due to poisoning by the substance dicoumarol which is present in the moulded bay and' which acts as an anti-vitamin K. Deficiency of the vitamin has not been recorded as a cause of losses in Britain.

The Water-soluble Vitamins : *Vitamin C* is required only by guinea-pigs, humans, and the higher apes, where a lack causes scurvy. Farm animals and birds manufacture the vitamin in their systems.

Vitamin B Complex - The remaining water-soluble vitamins are all grouped under this one name, and in the last twenty years the number of members of this group has grown to twelve. There are probably others yet undiscovered. No member of the vitamin B complex is required by adult ruminants, the reason being that the micro-organisms of the rumen, besides transforming food proteins and breaking down the fibrous constituents of feeds, manufacture the vitamins of the B complex in sufficient amounts to meet the need of their host. Before the rumen of the young animal is fully established i.e. while it is eating no solid food it needs a source of these B complex vitamins in its ration, but once the rumen flora is established the need comes to an end. Recent experiments have shown that horses do require some members of the vitamin B complex, but it does not appear likely that practical rations would ever be deficient. Hence, pigs and poultry, and the farm dog, are the only farm animals which are ever likely to suffer from major deficiencies.

Some of the individual members of the vitamin B complex are given in the table opposite, which summarises all that was known, at the time of writing, about their essentiality in the rations of pigs and poultry.

Tlie Vitamin B Complex

Vitamin	Other Names	Cheif Signs of Deficiency as observed in Experiments		Agricultural Importance
		Pigs	Poultry	
Aneurin	B_1	Loss of weight	Nervous disease (polyneuritis) causing abnomal posture	Very little importance, Since cercals and millers' offals are rich in the vitamin
Riboflavin	B_2	Limb stiffness	Curled toe paralysis of chick	Of importance for baby chicks which should receive some riboflavin supplement, such as dried whey or dried skimmed milk if given simplified rations
Nicotinic acid	PP factor (Pellagra preventive)	Severe diarrhoea and loss of appetite	Inflammation of the mouth and darkening of the tongue	Deficiency has occurred in farm practice where rations are restricted to maize products; but lack of the vitamin unlikly in normal practice
Pyridoxine	B_6	Anaemia and fits	Convulsions	Deficiency in ordinary rations unlikely
Pantothenic acid	...	Nervous disturbances and "goose-stepping"	Failure of growth	Very unlikely to be absent from common farm rations
Biotin	...	Cracked hooves	Poor growth and skin disorders	Very unlikely to be deficient in farm rations
Choline	...	...	Hock disorder called "perosis"	Unlikely to be deficient in farm rations
Folic acid	...	Anaemia	Anaemia and poor feather development	Very unlikely to be deficient in farm rations
B_{12}	...	Anaemia and poor growth when restricted to rations containig vegetable proteins only	Anaemia and poor growth when restricted to rations containing vegetable proteins only	Of imprtance as a constituent of the animal protein factor

These requirements have been established by experiments in which highly artificial rations were fed-composed of highly purified feeds, such as raw starch and glucose. These experiments, while they have established that the pig and chick require these members of the vitamin B complex, give no indication of their importance under practical conditions of husbandry. In general, deficiencies of vitamins of the B complex in pigs and poultry are rare, though "curled-toe paralysis" and "perosis" in poultry do appear occasionally. The former is due to a deficiency of riboflavin in the ration, and for best results starting mashes for chicks should contain 5 to 10 per cent, dried whey, or dried skim-milk powder, which are rich in riboflavin.

Considerable attention has been paid recently to the so-called "animal protein factor." At the date of writing it appeared that this was not a single "A.P.F." but several distinct factors, including vitamin B_{12}' "A.P.F." is found associated with protein in feeding stuffs derived from animals but is absent from those derived from plants. It appears that the poor growth of pigs and poultry restricted to rations containing only plant proteins is in part due to deficiency of this factor, and that the addition of A.P.F. concentrates makes such rations as good as those containing fish meal, meat meal, or milk. In Britain, as the result of long experience, it has become a usual practice to incorporate fish meal into most rations for rapidly growing pigs and poultry; hence deficiencies of A.P.F. arise only infrequently.

Closely associated with the A.P.F. is the discovery that antibiotics, such as penicillin and aureomycin promote growth. The increase in live-weight gain may result from the control, by the antibiotic, of very slight and perhaps imperceptible "infection" of the animals which respond. This "infection" has been transmitted from bird to bird but its nature at the time of writing was unknown.

Food Requirements : From what has been said it will be clear that animal nutrition is a very complex science. An animal requires in its food some eight or ten essential amino-acids, about fifteen vitamins, and about the same number of essential mineral elements, as well as carbohydrates and fats as sources of energy. Ruminants, by virtue of the activity of their rumen flora, can dispense with some of these nutrients.

Of all these many requirements the one which is of primary importance is energy. It has been estimated that by far the main part of the economic loss that is due to faulty nutrition of farm stock results from failure to meet the animals energy requirements. These are usually divided into maintenance and production requirements. The requirement for maintenance is the amount that the animal must expend on the various processes and reactions of the body which are essential to life. An animal is, thus, very unlike an internal-combustion engine. When the latter is not being used it does not require fuel. When a milking cow is dry or when a steer is not growing, or fattening, it still requires a source of energy to keep it "ticking over." The production requirement is the amount of energy which the animal needs, in addition to its maintenance demand, to support the production of milk, eggs, meat, or fat, or to meet the energy lost to the body in muscular work.

Energy requirements are measured in terms of Calories, one Calorie being the quantity of heat energy which is required to raise the temperature of 1 kg. of water 1°C. It is written with a capital C, or called the kilogram Calorie, to distinguish it from the "calorie," 1000 of which make up the kilogram Calorie. A thousand kilogram Calories make one therm.

The total or *gross energy* value of a feeding stuff, of a particular nutrient or, indeed, of a fuel, is expressed as the amount of heat which it produces when it is burnt completely, the carbon and hydrogen it contains being completely oxidized to carbon dioxide and water, and the nitrogen liberated as free nitrogen gas. The estimation is made in an apparatus known as a bomb calorimeter. A known amount of the material is placed in a steel bomb in an atmosphere of pure oxygen, and the bomb is placed in a known volume of water. The bomb is then fired electrically, the material burns, and the rise in temperature of the water is noted. The amount of heat given out can then be calculated. When pure fat is burned its calorific value is found to be 4.2 therms per lb. Pure starch gives a value of 1.86 therms, linseed cake 2.1 therms, and hays, straws, and cereal meals about 1.8 therms per lb. Anthracite coal gives a value of about 3.6 therms per lb. It is obvious that these figures do not give very much indication of the value of the food to the animal, otherwise we would conclude that anthracite coal was a better feed than linseed cake, and that barley meal was about equal in feeding value to wheat straw. Gross energy values only state what energy the food contains; they give no information at all on the amount which the animal can obtain from it.

In the first place, only such portions of the food which can be digested by the animal furnish it with any useful energy. Those portions which appear in the faeces have not been digested. Furthermore, some substances are excreted by the body before they have been completely oxidized. Thus, the nitrogen contained in the digestible protein of the food is not excreted as gaseous nitrogen, but is locked up in compounds, such as urea, hippuric acid, and ammonia. These substances are incompletely oxidized and their excretion means a loss of energy to the animal. Again, the fermentations which take place in the rumen of the ox or sheep entail the production of combustible gases, mainly methane or marsh gas, and, under some conditions, hydrogen. This again represents a loss of energy to the animal. To obtain a measure of the useful energy which an animal obtains from its food it is necessary to deduct from the gross energy which it ingests the energy lost in the faeces, the urine, and as combustible gas. The balance is called *metabolizable energy,* a term which is better than "useful" energy since, as will be explained later, not all of it can be used by the animal for meeting its requirements for maintenance or production.

Professor Arms by, an American who was one of the first to study the energy metabolism of farm animals, gave the following example of the calculation of the metabolizable energy of a steer's ration. The steer was given a ration of timothy hay and its liquid, solid, and gaseous excreta were collected. The total energy values of food and its excreta were then determined :

	Calorific value (therms)
Energy of food	
6988 gm timonthy hay	27.72
400 gm linseed meal	1.81
Total intake	29.53
Energy of excreta	
16619 gm faece	14.23
4357 gm urine	1.21
142 gm methane	1.90
Total excreated	17.34
Difference, being the metabolizable energy of the ration	12.19

Not all the metabolization energy, however, can be used by the animal for productive processes or for meeting its maintenance requirement. A portion is always lost as sensible heat. This is not surprising since, to take an example in the manufacture of milk by the cow, it is clear that a great deal of internal work has to be done. The cow must spend part of the metabollzable energy in the work of prehension, mastication, digestion, and assimilation of her food, and a further part in the transference of absorbed nutrients to her udder. Even there, the efficiency of the biochemical transformations involved in the making of milk is never a hundred per cent., so that there is a further wastage of dietary energy. The energy expended in the work of digestion and the losses entailed in metabolic transformations of metabolizable energy is eventually given up as heat. This heat may be of value, under certain conditions, for keeping the animal warm, but in general it is a waste product, and, indeed, in very hot weather, is a liability.

For each animal receiving a particular ration there is what is called a critical temperature. This represents the air temperature at which the animal's normal body processes generate just sufficient heat to maintain its body temperature at the normal level. If the air tempera-ture drops below this critical temperature, the animal has to expend energy in movement, such as shivering, in order to keep its body temperature at the normal level. When chicks huddle together or pigs crawl under their litter, it may be assumed that the air temperature has dropped below their critical temperature for the prevailing conditions of feeding, humidity, etc. With the adult pig on full feed the critical temperature is about 40°F; only if the temperature falls below this level will food energy be diverted from the useful path of laying down fat and protein in the body and be squandered by the animal in order merely to keep itself warm. Young pigs and baby chicks have much higher critical temperatures than adult pigs and hens, and thus, it is important, when the weather is cold, to provide them with additional heat. Ruminants, on the other hand, when given normal rations, have very low critical temperatures. Cowshed temperatures below freezing-point have no effect on the utilization of feed by cows. This is largely to be explained by the fact that the extensive fermentations which take place in the paunch produce a considerable amount of heat; thus, while a pig derives about 1500 kilogram Calories from I lb. of barley meal, a cow or sheep loses, as heat, about 500 extra kilogram Calories. It is, thus, a mistake to suppose that feed can always be saved by keeping animals warm. This will happen only if the temperature of the surroundings drops to below the critical temperature, and is likely to occur only with pigs or poultry when poorly housed. It is true that, in severe winters, deaths of hill sheep do occur, but the deaths are largely due to the combination of starvation and cold; as the animals get less and less to eat, and thus, have less and less waste heat available, the critical temperature rises. A fasting animal has a critical temperature which may be as much as 50° F. above that of one on a normal diet. Thus, the critical temperature of the hill ewe rises, her reserves of energy in the form of body fat are exhausted, and eventually she can no longer equate her heat production with heat losses, when the body temperature falls and dies.

At the other end of the scale high air temperatures impose a considerable strain on the animals' means for getting rid of heat. At high environmental temperatures, cattle and sheep, which have no well-developed sweating mechanism, have to rely on the evaporation of water from their respiratory passages to keep themselves cool, and pant vigorously. As air temperature rises further the animal is unable to get rid of enough heat and the body temperature rises. This rise appears to result in the acceleration of many of the body processes, and the animal's expenditure of energy, thus, increases. The temperature at which this takes place is called the *point of hypothermal rise,* and between this and the critical is what is called the *range of thermal neutrality.* Within this range the animal's expenditure of energy is not affected by variation of air temperature. Just as a lowered food intake increases the critical temperature so, at the other end of the scale, a lowered intake raises the point of hypothermal rise. In other words, an animal on a low level of nutrition has a relatively high heat tolerance. This is of considerable importance in tropical countries, since it means that, under the hot, humid conditions, high levels of feeding cannot be used. Dairy cows imported from temperate regions deteriorate under these hot conditions and give little milk because they cannot maintain the food intake necessary for high milk yield and yet keep their body temperature normal. The tropical breeds of cattle are better equipped to get rid of surplus heat than are those of temperate regions.

If from the metabolizable energy we deduct the amount of energy lost in the form of heat as a result of (1) fermentation in the alimentary tract, (2) the work of digestion, and (3) the inefficiency of metabolic transferences within the body, we obtain a balance which is known as *net energy*. The net energy per pound of food ingested is known as the *net energy value* of the food concerned. It represents the energy of the feed which the animal actually uses for maintenance of its vital processes or deposits as fat or protein in its body or secretes in its milk. The maintenance cost, in terms of net energy is, however, a first charge on the total supply. Only the surplus is available for tissue growth, fat deposition, or the production of milk or eggs.

The net energy values of some feeding stuffs and some pure nutrients are given in the following Table :

Energy Values in Therms per lb. as determined with Fattening Cattle

Food	Gross Energy	Energy lost in Faeces	Urine	Methane	Metabolizable Energy	Energy Lost as Heat	Net Energy Value
Earth-nut oil	3.99	0.00	0.00	0.00	3.99	1.74	2.25
Wheat gluten	2.63	0.00	0.49	0.00	2.14	1.18	0.96
Starch	1.86	0.00	0.00	0.19	1.67	0.69	0.98
Maize meal	1.80	0.21	0.08	0.16	1.35	0.52	0.83
Timothy hay	1.81	0.86	0.07	0.14	0.74	0.31	0.43
Wheat straw	1.85	1.08	0.04	0.15	0.58	0.47	0.11

* In more extensive experiments, Kellner found the average net energy value of starch to be 1.071 therms per lb.

It is clear from the table that the proportion of the gross energy which is metabolizable, and also the proportion of the meta-bolizable energy which is really of value to the animal as net energy, varies considerably from one feeding stuff to another. Although the gross energy value of wheat straw is slightly higher than that of maize, the fattening steer obtains only about one-eighth as much net energy from straw as from maize.

The net energy values of feeding stuffs are not, however, the same for all animals. This was first shown by Oskar Kellner, a famous German nutritionist, who devised the system still used in Britain for the rationing of livestock, and which has since been extended by other workers. Kellner found that when he added 1 lb, of starch to the maintenance ration of a steer, the steer stored 0.25 lb. of body fat each day. Since the energy value of fat is about 4.28 therms per lb., the net energy value of starch for the steer was 1.07 therms per lb. When the same experiment was made with the pig, however, it was found that the net energy value was 1.50 therms per lb. The pig is, therefore, a much more efficient animal than the steer in utilizing the energy of materials, such as starch. This difference in efficiency reflects the differences between the digestive process of the two species. The pig has a simple digestive system well adapted to the utilization of concentrated foods, while the digestive process of cattle involves a massive bacterial fermentation of feeds prior to their digestion lower down the alimentary canal. This bacterial fermentation, though well-adapted to make the greatest use of coarse fibrous foods, necessarily wastes a good deal of the energy of easily digestible materials, such as starch. The difference in efficiency in the utilization of starch by the pig and steer does not mean the pig is, in general, superior as a food converter. The opposite conclusion would have been reached, if a feed containing 12 to 15 per cent, of crude fibre had been used in the experiment.

Since the net energy value of a feed is the difference between the metabolizable energy it supplies and the expenditure of energy involved in its assimilation, it will be clear that, if the expenditure of energy exceeds the supply, the net energy will be a negative quantity. Such values have been found with sawdust and peat moss. Wheat straw has very little value for ruminants, and for horses is worse than useless, since the horse expends far more energy in eating and digesting it than the straw yields in metabolizable form.

The basis of livestock rationing in Britain is the net energy system outlined above, but energy values are expressed in other units, viz., in terms of a reference substance, starch, rather than as kilogram Calories. The *starch equivalent* of a feeding stuff, thus, represents the number of pounds of pure starch which would have to be fed to an animal to give the same practical result in terms of fat deposition, as 100 lb. of the feeding stuff. The starch equivalent system was devised by Kellner,- and is essentially the basis of most Continental systems of rationing, such as the feed-unit systems of Scandinavian countries.

The determination of the starch equivalents of feeding stuffs is an extremely difficult task. Kellner determined them by experiments with fattening steers in an apparatus which enabled him to measure the storage, by a steer, of both carbon and nitrogen. This was done by confining the steer in a respiration chamber, measuring the carbon lost in dung and urine, and calculating the carbon-content of the carbon dioxide lost in respiration and also in the methane produced by fermentation. From the results Kellner calculated the amount of fat stored by the steer when a particular feed was given as an addition to a maintenance ration. He, thus, found that the addition to the maintenance ration of 10 lb. of linseed cake resulted in the formation of as much fat as was produced from 7.4 lb. of starch. Hence, expressed per 100 lb., the starch equivalent of seed cake is 74. In the early years of the present century Kellner worked out many starch equivalents, but later realized that a more simple method of determination was needed. He, therefore, devise a method for calculating the starch equivalent of a feeding stuff from its content of digestible nutrients, i.e. from the amount of digestible protein, digestible ether extract, and digestible carbohydrates it contained. The energy value of the above nutrients were found to be as follows:

1. Each pound of digested crude protein supplies as much energy as 0.94 lb. of starch.
2. Each pound of digested carbohydrates supplies the same amount of energy as I lb. of starch.
3. Each pound of digested ether extract supplies as much energy as 1.91 2.12, or 2.41 lb. of starch, the different values being allotted to allow for the varying composition of the ether extract of different feeds. The highest factor applies to oilseeds containing much true fat, and the lowest to hays and straws in which the ether extract is mainly pigments and resins. The intermediate figure applies to cereals.

When Kellner applied these factors to the digested nutrients of feeding stuffs and compared the calculated starch equivalent with his experimental observations made in the respiration chamber, he found that the results failed to agree. The calculated figure was always greater than that actually determined by calorimetry. The difference was greatest in the case of fibrous foods-hay, straw, and other coarse fodders-when the calculated figures were often as much as 40 per cent, too high. Obviously, the work of prehension and mastication of the fodder and of the propulsion of indigestible matter along the tract had not been taken into account in his calculations. Kellner, therefore, devised two sets of correction factors.

One he called "V factors," to be applied to meals and grains. These are simple percentage values indicating the discrepancy between calculated and observed results. Thus, a V factor of 97 for linseed cake means that the true starch equivalent of linseed cake is only 97 per cent, of the calculated value. The other corrections for roughages he based on a sliding scale to take into account the fibre-content, making a deduction from the calculated figure for each 1 per cent, of fibre in the material. For fresh green fodder the factor used was 0.29, and for the mature hays and straws 0.58. The following example shows how the starch equivalent of a sample of ryegrass hay may be arrived at:

Digestible protein, 4.9 per cent x 0.94	4.61
" fibre and nitrogen-free extract, 41.5 per cent x 1.00	41.50
" ether extract 1.4 % x 1.91	2.67
	48.78
Correction:	
fibre contents = 22.9 per cent x 0.58	13.28
starch equivalent per 100 lb	33.50

Tables in the Appendix give the starch equivalent, protein equivalent, and average content of digestible nutrients of most of the common feeding stuffs of the farm.

Most of the starch equivalents used in Britain for livestock feeding have been calculated in this manner; indeed, no single direct determination has been made with cattle. It must be remembered that the foods used by Kellner in Germany fifty years ago may have been rather widely different from those in Britain today. We have new crop varieties, and some new crops, such as sugar and fodder beets, new farm processes, such as grass-drying and modern ensilage. Again, the factory by-products now available for feeding, have been more highly processed than those of fifty years ago. It is, thus not surprising that many writers have criticized the starch-equivalent system, often because they have expected the calculated figures to be more precise than, in fact, they pretend to be. Calculated values are not precise, but the results of practical feeding trials do conform fairly closely to expectations based on the book values.

The starch equivalents calculated in the manner given above estimate the values of feeding stuffs to ruminants. As already pointed out, pigs and poultry will make better use than ruminants of some feeds. Armsby, for instance, found the net energy value of maize for fattening cattle to be 85 therms per 100 lb. and for the pig 120 therms. It would, of course, be useful to have a separate set of starch equivalents for pigs and poultry, but in general it has been found that the *relative* values of feeding stuffs are approximately the same for different classes of stock. Thus, experiments with pigs show the starch equivalent of potatoes to be 20 per cent, of that of maize, while experiments with cattle show potatoes to have 23 per cent, of the value of maize. Obviously hay and straw, which ruminants can and pigs and poultry cannot use, could not be expected to conform to this generalization, but, for all but fibrous foods, starch equivalents determined with cattle can be used in the computing of rations for other classes of farm stock.

Food Requirements : It is quite clear that starch equivalents are *estimates* of the amount of energy that animals obtain from particular feeding stuffs. The energy requirements of particular animals are estimates also and can be expected to apply only to the average individual. Two animals may be exactly of the same size, yet one may require more food than the other to maintain itself and make a given amount of growth or lay on a given amount of fat. This is understandable, for one may be a placid individual taking very little exercise and the other a highly strung creature dissipating much more energy in body movement or nervous tension. It must be emphasized that a system of rationing according to theoretical requirements is no more than an aid to good stockmanship. Precise rationing of animals according to feeding tables will, on average, give good results; but the knowledgeable stockman realizes the inevitable margin of error and makes minor adjustments in the ration of individual animals. Rationing, if done intelligently, prevents waste of food. For example, it was common farm practice, fifty years ago, to feed fattening bullocks enormous quantities of high-protein cakes. Much of this protein was surplus to the animals. Material needs, and the nitrogen it contained was simply excreted in tile urine. Knowledge of energy and protein requirements and values has enabled this waste to be avoided.

In rationing farm stock the primary object is so to adjust the energy and protein supply to enable the animal to realize something near its potential level of production and to avoid waste.

As said previously, mineral and vitamin requirements must also be provided for.

Energy Requirements for Maintenance : The maintenance energy requirement of an animal is made up of two parts: the basal or minimal metabolism and the activity allowance. Basal metabolism measures the heat lost by an animal when it receives no food at all and is lying in a state of complete rest. Very few determinations of it have been made with farm animals. Muscular activity increases the animal's energy loss-even standing without movement increases heat production by 15 per cent, while short bursts of activity, such as galloping or struggling increase it as much as three-fold or four-fold. In stall-fed cattle provision for basal metabolism normally accounts for about 70 per cent, of the total maintenance cost. It has been found by controlled experiments, and confirmed by feeding trials, that the average maintenance requirement (including both the basal allowance and that for activity) of a 1000-lb. ox under conditions of close confinement is 6 lb. starch equivalent per day. This value is only an average, and individual oxen weighing 1000 lb. may have maintenance requirements as much as 25 per cent, more or less. Obviously, on average, animals heavier than 1000 lb. will require more and lighter animals less. It has been shown by a careful study of the records made with many species that basal metabolism varies not directly with body weight but with body weight raised to the power 0.73. The live weight to the power of 0.73 is called "metabolic body size." It is a mathematical expression of the general finding that basal metabolism is roughly proportional to the animal's surface area, *i.e.* the area from which it radiates heat. This generalization applies, however, to mature animals only and to basal metabolism only. It does not permit any reliable estimates to be made, by simple arithmetic, of the maintenance requirements of very young animals. Young growing animals tend to have much higher maintenance requirements than would be estimated from their " metabolic body size." Thus, the maintenance energy requirement of a mature sheep weighing 100 lb. is approximately 1200 kilogram Calories per day, or 1.1 lb. starch equivalent per day, whereas a young calf of the same weight has a maintenance requirement of 1950 kilogram Calories. A young calf only one-tenth the weight of an adult cow has a maintenance energy requirement of about one-third that of the adult. With adult animals, however, it is fairly safe to use the metabolic body size in computing requirements. Even here, however, it appears that Jersey cows require more maintenance energy in proportion to their metabolic size than those of larger breeds. Indeed, it seems necessary to use slightly more liberal allowances for milking cattle than for others.

Maintenance Requirements of Growing and Fattening Cattle

Live Weight (lb.)	Maintenance Energy Requirement (lb. Starch Equivalent per Day)	Maintenance Protein Requirement (lb. Protein Equivalent per Day)
300	2.8	0.25
400	3.3	0.32
500	3.8	0.36
600	4.2	0.42
700	4.7	0.47
800	5.1	0.51
900	5.5	0.55
1000	6.0	0.60
1100	6.5	0.65
1200	6.9	0.69
1300	7.3	0.73
1400	7.7	0.77
1500	8.1	0.81
1600	8.5	0.85

Maintenance Requirements of adult milking cows of average size

Breed	Maintenance Energy Requirement (lb. Starch Equivalent per Day)	Maintenance Protein Requirement (lb. Protein Equivalent per Day)
Jeseys	5.5	0.15
Guernseys	6.3	0.55
Ayrshires	6.5	0.60
Shorthorns	7.4	0.70
Friesians	7.6	0.70

With sheep there are very few estimates of maintenance requirements, but recent experiments suggest that the older results are approximately correct. As with cattle, there are undoubtedly breed differences in maintenance requirement. Hill ewes doubtless expend more energy than a folded flock in searching for their food. The following table gives estimates of requirements for the maintenance of sheep :

Live Weight (lb.)	Maintenance Energy Requirement (lb. Starch Equivalent per Day)	Maintenance Protein Requirement (lb. Protein Equivalent per Day)
60	0.73	0.08
80	0.90	0.10
100	1.07	0.12
120	1.23	0.14
140	1.38	0.16
160	1.52	0.17
200	1.79	0.20

Under present-day conditions in Britain these figures are of no more than academic interest since the accurate rationing of energy for sheep, even on a group basis, is impracticable. In moving sheep from one pasture field to another, or in setting off an area of forage crop for, say, a week's consumption, or again in deciding whether or how much supplementary food should be offered, the farmer or the shepherd must be guided by experience, and by the appetite, bodily condition, and general behaviour of the sheep. It is only where sheep are housed in winter, or fed in "dry lot," that any system of rationing can be applied.

With pigs there have been many experiments, notably at Cambridge, in Denmark, and in Germany, in which maintenance requirements have been determined in calorimeters and respiration chambers. Their practical value is not very "great since most pigs grow too rapidly to allow adjustment of their rations on the basis of calculated maintenance requirements. The all-meal system aims to provide the pig with a ration containing about 70 lb., starch equivalent per 100 lb. of dry matter, and the daily allowance of meal is periodically adjusted to appetite. In the later stages of fattening for bacon some restriction of feed is often necessary to the production of good carcass quality, without excessive fat. With the Lehmann feeding system, the rationing is based on a restricted ration of meal containing sufficient protein and minerals to "balance" a bulky supplement of potatoes, fodder beet, kitchen waste, etc., which is fed to appetite.

With farm horses the basal metabolism is roughly the same as for cattle of equivalent weight, and the requirement for normal activity, when the horse is not working, is also about the same. The horse tends to make slightly better use of its feed, for maintenance, than does the ox, largely because fermentative losses are not so great.

Protein Requirements for Maintenance : The requirement of animals for protein is largely accounted for by their need for amino-acids to manufacture the proteins of new tissues or of milk. But animals that are neither growing nor producing milk need protein to replace the nitrogen-containing constituents of the tissues, which are constantly being broken down. An animal on a protein-free ration must, thus, lose nitrogen from its body. On such a ration nitrogen is found in the faeces, partly in the residues of digestive juices, and partly in the millions of cells that are abraded day by day from the walls of the digestive tract. Similarly, losses occur in the form of scurf and hair. The nitrogen in the urine represents the inevitable loss of cell constituents incidental to the normal metabolic processes of the body cells. Maintenance requirements of protein are measured by feeding animals protein-free or protein-low rations and measuring the total losses of nitrogen from the body. The figure so obtained represents the minimal requirement of the animal which, for an adult steer weighing 1000 lb., amounts to about 0.20 lb. protein per day. But the amount of food protein required to achieve balance is considerably higher, viz. 0.5 to 0.6 lb. of protein equivalent per day. As in the case of basal energy requirements, protein requirements for maintenance are proportionate to the metabolic body size and not directly to the body weight. By contrast with the maintenance energy requirement, however, muscular activity has no effect on maintenance protein requirements provided that sufficient energy is provided in the form of carbohydrate, fat, etc. The tables on pages 62 and 63 give the maintenance requirements in terms of protein equivalent of cattle, horses, and sheep. Obviously, the production of wool by the sheep-containing about some 2 or 3 oz. of pure protein per week-necessitates an additional requirement. Allowance is made in the tables for this extra demand.

Production Requirements : (1) *Energy : By* definition, 1000 kilogram Calories of net energy (0.94 lb. starch equivalent) fed to a fattening ox will result in the deposition of 1000 kilogram Calories in the form of fat. This is equivalent to about 4 oz, of pure fatty tissue, The production of 1 lb. of fat in a mature steer, thus, requires nearly 4 lb. starch equivalent. Fat is not only stored in fat "deposits," *e.g.* the subcutaneous layer and the "caul," but also in the muscles, and without any concomitant storage of water. Hence, the deposition of I lb. of fat means also I lb. of live-weight increase. But in animals that are both growing and fattening it is only in the last stages of fattening that I lb. of live-weight gain represents the gain of 1 lb. of fat. In growing animals part of the energy of the food that is surplus to maintenance is used in the building up of body tissues. Moreover, when protein is laid down in muscles and other tissues it is accompanied by about four times its weight of water. Thus, we find that, in young calves, the live-weight gain contains only some 10 to 15 per cent, of fat, the remainder being distributed between ash (about 10 per cent) and protein (15 per cent), with as much as 65 per cent, of water. The energy value of I lb. of live-weight gain, thus, varies widely-between about 800 kilogram Calories in animals growing and laying down muscle and bone, and 4000 kilogram Calories in fully grown fattening animals. Indeed, the energy value of T lb. of gain may be actually higher than the value of 4000 kilogram Calories since, in the last stages of fattening, added that may replace body protein or body water. Hence, the amount of starch equivalent required to produce I lb. of live-weight gain will vary from less than 1 lb. to fully 4 lb. A rough estimate of the requirement per pound of live-weight gain can be based on the age of the animal and the rate of live weight gain. Thus, a young calf, growing but not fattening, has the lowest requirement. An older animal, being maintained in store condition, will need rather more. A fattening beast requires more in the later than in the early stages of fattening. In the tables below account is taken of all these variables.

Requirements of Ruminants for Growth and Fattening

Normal growth without fattening (growing dairy stock and lambs destined for breeding)

CATTLE		SHEEP	
Age (months)	**lb. S.E. per L.b. Live-weight Gain**	**Age (months)**	**lb. S.E. per lb. Live-weight Gain**
0 to 3	1½	0 to 3	1¼
3 to 6	1½	3 '' 6	1¾
6 to 12	2	6 '' 12	2¼
12 to 24	2¼		

Fattering of young cattle and sheep (baby beeves, fattening lambs, and young fattering tegs)

Early stage of fattering..... 2 lb. S.E. per lb. lives weight

Later stage of fattering..... 3 lb. S.E. per lb. lives weight

Fattering of older cattle and sheep (cattle of 2½ years or over, including old cows: sheep over 9 months, including old ewes)

Early stages........2½ lb. S.E per l.b lives weight.

Middle....... 3 lb. S.E per lb. lives weight

Late.........3½ lb. S.E per lb. lives weight.

With pigs the same general principles apply but, as mentioned earlier, it is usual to express requirements in terms of total meal per day. Requirements are discussed in Chapter VI.

Again, in the case of milk it will be obvious that account must be taken of the variable composition of the product: for example, more food energy will be needed to produce a milk containing 5 per cent, of fat than one containing only 3 per cent. A gallon of milk with 4 per cent, of fat has a calorific value of 3400 kilogram Calories, which would suggest a requirement of about 3 lb. starch equivalent. In fact, the need is less because the conversion of energy into milk is more efficient than conversion into live weight gain. The usual recommendation is 2.5 lb. starch equivalent for each gallon of average milk containing 3.75 per cent, fat. But practical experience and also experimental evidence indicate that somewhat higher amounts lead to better sustained yields. The amounts of starch equivalent given below are 0.2 lb. higher than those usually recommended.

Energy and Protein. Requirements for Milk Production

Fat-content of milk (per cent.)	**lb. Starch Equivalent (per gallon)**	**lb. Protein Equivalent (per gallon)**
3.00 to 3.50	2.4	0.48
3.50 to 4.00	2.7	0.55
4.00 to 4.50	3.0	0.62
4.50 to 5.00	3.2	0.69
5.00 to 5.50	3.5	0.76
5.50 to 6.00	3.7	0.83

Even so, higher intakes of starch equivalent than the standards above suggested will result in higher milk production. Much recent experimental work has shown that high planes of nutrition, entailing feeding at levels above those given by feeding standards, will result in substantial increases in milk yield and, incidentally, in body weight. Many of the high average yields of individual dairy herds may be traced to the very high levels of feeding which are employed. High levels of feeding may not, indeed, always be profitable. In general, however, it should be remembered that levels of feeding above the accepted standards are bound to give some increase in milk yield since the law of diminishing returns applies to milk production; but experience has shown that it may take nine months, or perhaps longer, before the effects of a large increase in the feeding level are reflected in an overall economic gain of milk.

For work production there are no reliable figures, and the following are rough guides to feeding working horses :

Requirements for Working, Horses

	Starch Equivalent Required per Working Hour (per 1000 lb. body weight)
Light slow work	¾
Medium slow work	1
Heavy slow work	1¼
Heavy fast work	1½

(2) *Protein :* The amount of protein required for productive purposes is determined by much the same factors that apply in the case of energy. If the live-weight gain is largely protein, as it is in the case of a young and rapidly growing calf, then relatively more protein will be needed than in that of an adult fattening beast, which is storing up little except fat. As the young animal grows, it, of course, has a rising maintenance requirement, but proportionately less for the build-up of tissue. These two opposites almost cancel out, so that, for instance in the case of cattle, a constant daily allowance of 0.9 lb. protein equivalent per head will cover the daily requirement from an age of about five or six months (3 cwt. live weight) to maturity. A constant allowance of 0.9 lb. protein equivalent over this weight range, having regard to rising energy needs, thus, implies a widening of the nutritive 'ratio of the ration as the animal gets older. The nutritive ratio may be expressed as that between protein equivalent and the starch equivalent in the ration as a whole. The same figure applies to

fattening cattle, good results having been obtained on a daily allowance of 0.9 lb. In practice, however, more than this amount of protein is ordinarily given, because rations with very wide nutritive ratios are rather unpalatable. The figure of 0.9 lb. should, thus, be regarded as a minimum. A higher allowance will do no harm, and although it may increase the cost of the ration it will often be necessary to maintain the desired level of food consumption. An allowance of less than 0.9 lb. may retard growth.

With sheep and lambs the same general principles apply, and a constant allowance of 0.20 lb. protein equivalent per head per day will suffice for maintenance, growth, and fattening in lambs and tegs of ordinary size.

Turning to the dairy cow, it has been shown that milk of high fat-content also has a high protein-content. Hence, the protein allowance, per gallon, must vary with the fat-content of the milk. In the past it has been customary to feed 0.6 lb. protein equivalent per gallon of average composition; but extensive feeding trials have shown that this figure can be reduced to 0.5 lb. without causing any significant fall in yield. These results may indeed reflect the "margin of safety" in the maintenance part of the ration; and despite the experimental evidence for the level of 0.5 lb. it may be better, in practice, to adopt a standard of 0.55 lb. per gal. Protein requirements for milk production are set out in the table on page 121.

Appetite : There is a fairly definite limit to the amount of food that an animal can eat in a day or at a meal. On the other hand, there is a minimum required to satisfy the appetite. Too bulky a ration will simply be refused, while a too concentrated one, besides not being sufficient to satisfy the animal's hunger, may result in digestive disturbances. The bulk of a ration is usually measured by its dry-matter content, but this measure is not entirely satisfactory. Animals will, in general, eat more of high-quality foods than of low-quality fibrous ones. It is probable that "space-tilling" is better measured by the content of indigestible residues, or "ballast," than by dry-matter content. The former measure is used on the Continent, but it is not the ideal one. In this country we commonly express appetite, and the need for bulk, in terms of dry matter, recognizing, however, the limitations of this measure. For cattle and horses the lower limit is about 15 lb. dry matter per 1000 lb. live weight daily, while normal rations may, in general, contain upto 25 lb. Dairy cows tend to have larger appetites than other classes of cattle; in fact, some very heavy milkers can consume as much as 40 lb. dry matter daily if the total amount is divided into four or five meals. Lean store stock can also deal with slightly more than the usual 3 lb. of dry matter per cwt. of live weight, which is the commonly accepted figure in their case. For sheep 2½ to 3 lb. dry matter per 100 lb. live weight per day is a normal intake. With pigs a good ride is to arrange for the starch equivalent of the ration to be three-quarters of the dry-matter content. Pigs will eat about 4½ lb. dry matter per day per 100 lb. body weight when weaned at two months old, decreasing to about 3 lb. as they reach bacon weight.

In general, appetite is expressed per 100 lb. of live weight. But the figures quoted above do not apply to the young calf or to other animals at the suckling stage. These must have a diet of high energy value in relation to its bulk, *i.e.* composed largely of concentrated and easily digested materials, if full growth is to be secured.

There are many other factors which must be taken into account in the practical rationing of animals. The palatability of foods is one which is often ignored but is of considerable importance. When strange foods are introduced they generally prove unpalatable to at least some individuals. Certain breeds, such as Jerseys, tend to be more than ordinarily fastidious. The effect of the feed on the health of the animal is another matter. Unsterilized bone flour might possibly introduce anthrax spores and cause disease, while there is much farming experience to suggest that some feeds are conducive to fertility or, conversely, result in poor reproductive performance. The latter opinion, however, has not been sufficiently examined experimentally. Some foods, such as rice meal and maize, result in the production of too soft a fat in bacon pigs, while others produce a milk fat that is too hard or too soft for butter-making. Other feeds, such as turnips and molassed sugar-beet pulp, may cause taints in milk. Again, some feeds, such as cotton cake, have a costive action and thus, may find a use at particular times 'when the major part of the ration is too laxative. Conversely, linseed cake is laxative.

Hence, in drawing up ration, it is not enough to adjust the starch equivalent and protein equivalent to the calculated requirements. Individual animals do not conform closely to general rules.

The following examples illustrate the application principles set out above :

1. *Daily Winter Ration of a Rapidly Growing Dairy Heifer Twelve Months Old weighing* 500 lb : The appetite of this animal will be about 14.5 lb. dry matter. For maintenance it will require 3.8 lb. starch equivalent. Normal growth will be about 14.5 lb. daily, implying a further 3 lb. starch equivalent and giving a total of 6.8 lb. Protein requirements will be 0.9 lb. If mangolds, hay, and dredge corn (equal parts oats and beans) are available, the calculation would be:

Energy supply	L.b.S.E
20 lb. mangolds (starch equivalent = 6.2)	1.24
12 good hay (starch equivalent = 37)	4.44
1 lb. oats (starch equivalent = 60	0.60
1 lb. beans (starch equivalent = 66)	0.66
Total	6.94
Protein supply	L.b. P.E
20 lb. mangolds (protein equivalent = 0.4)	0.08
12 lb. hay (protein equivalent = 4.7)	0.56
1 lb. oats (protein equivalent = 7.6)	0.08
1 lb. beans (protein equivalent = 19.7	0.20
Total	0.92

The total dry weight of the ration can be calculated to be 14½ lb. daily. The above ration could then be fed for a period and later be adjusted according to the results secured.

2. *Ration for a, Friesian Cow giving 4 Gals, of Milk containing* 3.75 *per cent. Fat. The* requirements of the cow would be:

	Starch Equivalent (lb. per day)	Protein Equivalent (lb. per day)
Maintenance for 1250 lb. live weight	7.7	0.70
Production (2.7 lb. starch equivalent and 0.55 lb. protein equivalent per gallon)	10.8	2.20
	18.4	2.94

The appetite of this cow will probably be about 3 lb. of dry matter per cwt., *i.e.* 33 lb. daily. Under present-day conditions it is commonly advisable to feed bulky home-grown foods for part of the production ration as well as the maintenance ration suitable ration would be :

Energy supply	lb. S.E.
20 lb. marrow-stem kale (starch equivalent 9)	1.80
20 lb. arable silage (starch equivalent 12)	2.40
15 lb. good meadow hay (starch equivalent 37)	5.50
6 lb. crushed oats (starch equivalent 60)	3.60
4 lb. cracked beans (starch equivalent 66)	2.64
2 lb. dried sugar-beet pulp (starch equivalent 61)	1.22
½ lb. decorticated ground-nut meal (protein equivalent 73)	0.36
1 lb. flaked maize (starch equivalent 84)	0.84
	18.41

Protein supply	lb. P.E
20 lb. marrow-stem kale (protein equivalent 1.3)	0.26
20 lb. arable silage (protein equivalent 2.0)	0.40
15 lb. good meadow hay (protein equivalent 4.6)	0.69
6 lb. crushed oats (protein equivalent 7.6)	0.45
4 lb. cracked beans (protein equivalent 19.7)	0.79
2 lb. dried sugar-beet pulp (protein equivalent 5)	0.10
½ lb. decorticated ground-nut meal (protein equivalent 41)	0.20
1 lb. flaked maize (protein equivalent 9)	0.09
	2.98

In practice the purchased sugar-beet pulp, decorticated groundnut meal (or cake), and flaked maize would be mixed with the home-grown oats and beans, 2 per cent, of a mineral mixture added, and the whole fed at the rate of 4 lb. per gal. for milk produced above the first gallon daily. The ration of marrow-stem kale, silage, and hay would be sufficient for cows producing one gallon only. The physical condition of the concentrate mixture might be rather fluffy, a difficulty that could be overcome by adding some heavier material, such as maize gluten feed or palm kernel cake; both of these are approximately balanced for milk production.

CHAPTER III

CATTLE-TYPES AND BREEDS

THE term 'cattle', originally used to include all classes of livestock, is now generally restricted to oxen, which are included by zoologists in the genus *Bos.* This genus comprises the wild and domesticated buffaloes of Asia, and some two or three wild African species of buffalo; the American and the European Bison; the Yak of Tibet; three species of Asiatic cattle-the Gaur, Gayal, and Banteng-all of which are found wild, while the last two are also kept under domestication; the domesticated lumped cattle (Zebus) of Asia and Africa; and lastly our European domesticated types. The buffaloes do not cross with the others in the list, but these others cross freely, and the hybrids (or at least the female hybrids) are fertile.

Domesticated cattle are known to have existed in Babylon as early as 5000 B.C. The earliest known domesticated type found in Europe is that known as *Bos taurus brachyceros* or *Bos longifrons,* a very small, slightly built ox with short horns, which was very widespread in the earliest Neolithic (polished stone) period. There is no evidence that it was evolved in Europe, and the accepted view is that it was introduced from Asia as an already long-domesticated animal. This type, sometimes known as the " Celtic Shorthorn," persisted in Britain till Roman times, but on the Continent it was displaced much earlier by a large, strong boned, and long-horned animal known as *Bos taurus primigenius.* This seems to have been produced either by crossing the older form with the native wild species, or perhaps by direct domestication of the latter, the old type being meanwhile discarded. The European wild Ox or Drus *(Bos primigenius)* was common in Europe during the Roman period, and the last known specimen was killed in Poland in 1627. It was a very large and powerfully built animal with long horns, and probably brown in colour.

European cattle comprise a great variety of types. Quasi-zoological classifications of these types have been proposed, but none is very satisfactory. As regards the British breeds, they are of very varied and frequently of mixed ancestry. Thus, the native cattle before the Roman invasion were small, black or brown in colour, with short horns *(Bos brachyceros).* The Jersey, Kerry, and Shetland probably show the largest proportion of this blood. Then, according to Professor Wilson, the Romans brought in large white animals whose descendants have been preserved as the "wild" white cattle of Cadzow, Vaynol, and Chartley, now known as Park Cattle. According to another view these park cattle are the direct descendants of the wild Urus. It is certain that the Anglo-Saxons brought over cattle which were probably red in colour and from which the Sussex and the Devon are presumably descended. Possibly the Norse invasions gave us our polled breeds, though occasional polled animals occurred before that time. Much more recently, in the seventeenth century particularly, there was a large influx of large, short-horned, and broken-coloured cattle from the Netherlands. These have had a large influence on the Shorthorn and Ayrshire, and probably a lesser influence on several other of our breeds.[1] These types of cattle have intermixed to a considerable extent, just as Celt and Saxon, Norman and Dane have intermixed to form our human population.

For practical purposes we classify cattle according to the object or objects for which they are kept, *i.e.* for labour, beef, milk, or any two, or all three. In Britain, oxen are scarcely used for work purposes, and therefore, our classification is into beef, dual-purpose, and dairy types. As a matter of fact there is a gradation of types from the Jersey, which is of practically no value as a beef animal, to certain strains of Shorthorns, Herefords, and other breeds which have been selected almost entirely on beef points and which produce only enough milk to rear their calves (perhaps 200 to 300 gals.) or sometimes less. Specialisation has been carried further in Britain and America than in most other countries. On the Continent of Europe a considerable majority of breeds may be said to be dual-purpose or triple-purpose, and purely beef breeds are few and relatively unimportant.

We must endeavour to analyse as far as possible the qualities that constitute merit in a beef or in a milking animal. The breeder's ideal is, of course, not simply a high quality of meat or a large yield of milk, but beef or milk or both in such quantity and of such quality as will leave a maximum of profit to the producer. Our aim in breeding must, therefore, depend on the demands of the consumer, and it is of course true that these may

[1] See Wilson's Evolution of British Cattle, Vinton, 1909.

change from time to time. An example of such change is to be seen in the demand for small, tender joints, with a high proportion of lean meat, as compared with the popular taste of a century ago for larger joints of more mature and very much fatter meat. Nevertheless the main characteristics of a good beef or dairy animal have a permanent significance.

It follows from what has been said in Chapter II of this section that the efficiency of an animal, whether in the production of meat, of wool, of work, or of milk, depends on the ratio that obtains between the total food consumed and the quantity required for maintenance. The animal is always somewhat inefficient in the sense that it must consume a certain amount of material before it begins to produce anything at all. An animal that eats no more than a maintenance ration has obviously an efficiency of zero, and efficiency rises as production increases. Efficiency may be expressed as the energy value of the product stated as a percentage of the energy value of the food digested. On this basis a cow of medium size yielding 750 gals. per annum has an efficiency of 25 per cent. With a yield of 1100 gals. the efficiency rises to 35 per cent. Thus, the more our animal eats in proportion to its maintenance ration (so long as the food is efficiently converted into the required product and the animal remains in health) the more economical does the productive process become. Hence, whether for beef or milk-production, we want a low maintenance requirement and a capacity to deal with large amount of food. The maintenance requirement in cattle at rest has been found to be fairly uniform in proportion to the animal's size but in practice the non-productive consumption does vary within rather wide limits according to the temperament of the animal and according to its opportunities for rest. The energy expenditure is substantially higher (about 15 per cent.) when the animal is standing than when it is lying down; and substantially higher again, of course, when it is walking or running. A restless animal may waste a very considerable amount of energy in useless movement. A further point is that high milk production seems to be associated with a relatively high metabolic rate, whereas a lower level of metabolism is associated with rapid fattening.

We must next briefly consider the question *of early maturity,* which is very important in the breeding of all meat-producing types. The new-born animal is short-bodied, shallow-bodied, and long-legged. If such an animal be slaughtered and the different tissues of its body separated, it will be found that the proportion of bone and offal (heart, lungs, digestive organs, etc.) is very large, that of muscle tissue (lean meat) small, and that of fat very small indeed. As the animal grows the muscle tissue increases much more rapidly than the bone, and the general build becomes longer, deeper, wider, and more blocky. At a later stage still, if a full diet is provided, the proportion of fat rises; and when the accumulation of fat has reached the optimum (which depends on the consumer's taste) the animal is ready for slaughter. In the case of the sheep, Hammond found the proportion of bone, muscle, and fat in the leg of the new-born lamb to be 100 : 159 : 9, whereas in the mature fat sheep the ratio was 100: 560 : 285. Now the rate at which these changes occur varies from individual to individual and from breed to breed, and so we distinguish early maturing from late maturing types. It should be understood that early maturity has nothing to do with weight for age. It is rather the capacity to " telescope" the three phases of development which follow one after the other in the late-maturing animal *i.e.* (1) the growth of bone and offal, (2) the growth of muscle, and (3) the deposition of fat. Thus, a South Devon or a Friesian steer at eighteen months old would normally outweigh an Aberdeen-Angus; but the latter would normally have a much higher proportion of fat and muscle, and a lower proportion of bone, than the others. Another important point is that while the tendency to mature early or late is an inborn characteristic, the actual achievement of early maturity depends largely on feeding. An animal of the 'best beef breeding, if it be kept upon a very low ration, will remain narrow, leggy, and shallow-bodied; hence there is no object in developing early maturity in strains of animals that have to live through periods of severe scarcity. Indeed an early-maturing type of animal will be permanently stunted by restriction of its food intake to a level that would suffice for' the normal development of a slow-maturing type.

The object of breeding for early maturity is to achieve economy in production under favourable food conditions, because the shorter the period taken to produce a finished beast, the smaller is the proportion of the total food that is spent on maintenance. Moreover, the early-maturing animal can produce meat that is fat enough while it is still tender.

The other chief aim is quality in the product-the fat animal. This is to some extent a question of the percentage of meat in the animal as a whole. Beef is of far greater value, generally, than the other parts of the body, and hence, the type of animal required is light of bone and hide and offal and thickly fleshed. Some of these characters are, however, incompatible with other desirable qualities e.g. a light middle generally indicates a poor feeder. The proportion of the different cuts is also of importance. An animal with its weight concentrated in the back, sirloin, rump, and hind quarters, which are the high-priced cuts, will be much more valuable than another with a heavy neck and shoulder and a thin back. But here the differences that exist between different animals are not very large. A further point that determines the quality of the beef is the manner in which the fat is stored. Many unimproved breeds and dairy-bred animals tend to store the fat internally (for example, round the kidneys) and in a thick layer under the skin, the muscle tissues themselves remaining relatively free from fat. The result is that the carcass is wasteful, and that the lean meat, when cooked, is tough and dry. In good beef, on the other hand, much fat occurs within the muscles which, when cut across, show a characteristic "marbled" appearance, and the meat when cooked is tender, juicy, and of good flavour. This tendency to store fat abundantly within the lean meat is a most important quality in the beef animal. It is what the feeder means when he speaks of his animal "putting on flesh" during the fattening process. A minor point in connection with the quality of the meat is the colour of the fat. Most consumers object to a deep yellow colour, and some dislike a dead white. The colouring matter is largely carotene. Fat colour is in part a question of feeding and in part one of breed. The Channel Island breeds, for example, are characterized by the deep yellow colour of their body fat as well as of their milk. Mangolds and sugar-beet produce pale fats, while green foods, including green silage, give a deeper colour.

The points of a beef steer are shown in Fig. 7. The general type, which varies from breed to breed only in minor details, is a moderately long but deep and very wide block of meat on short legs. The head should be wide between the eyes and short from eye to nose; the muzzle large, the jaw strong, and the eye full and mild; the neck short, and full at its junction with the shoulder; the back wide, and almost uniform in width from crops to pin bones; the shoulder smoothly laid and well covered, neither protruding from the general outline of the side nor having a depression behind it; the brisket full, the heart-girth large, and the ribs

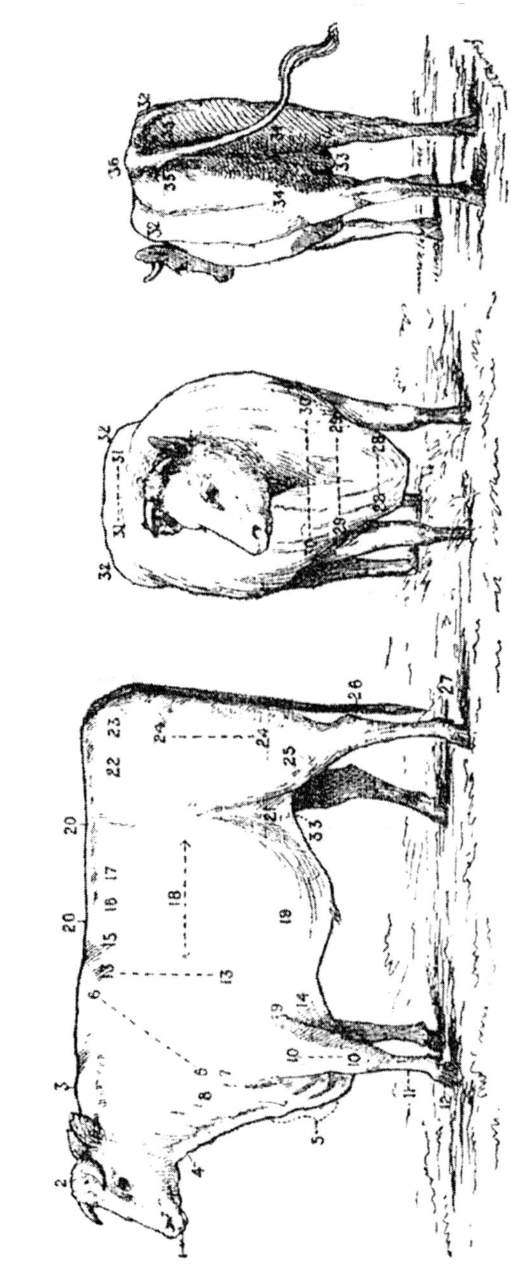

FIG. 33.—Illustrating the Points of an Ox (Thompson)

1. Muzzle	7. Shoulder point	13 to 13. Crop	19. Belly	25. Thigh	31 to 31. Loins
2. Poll	8. Shoulder or neck vein	14. Fore flank	20. Spine	26. Hock	32. Hooks or hip bones
3. Crest	9. Elbow	15. Fore ribs	21. Flank	27. Hind leg	33. Purse
4. Throat	10. Arm	16. Mid ribs	22 to 23. Plate	28 to 28. Brisket	34. Twist
5. Dewlap	11. Shank	17. Back ribs	23. Rump	29 to 29. Bosom	35. Pin bones or catch
6 to 6. Shoulder	12. Hoof	18. Barrel	24 to 24. Hip	30 to 30. Chest	36. Tail-head

well sprung or rounded, the middle moderately large, but not "paunchy," and the flank well let down i.e. the underline nearly straight. The animal should be well ribbed up *i.e.* the space from the last rib to the hook should be short; the hind quarters long, wide, level, and fairly square in cut, with the hook bones not too promi-nent; the plates well filled with flesh, the "tail-head" level and the tail hanging perpendicularly; the thighs thickly fleshed and the flesh well carried down to the hocks; the twist or inner thigh full. The animal should stand on short legs with fine but flat bone, and with the fore legs placed well apart. The flesh should be firm and resilient to the touch, not soft and flabby; the skin not too thin, but soft and mellow to the touch. "Handling" is very important as an indication of feeding capacity. In judging commercial stores the chief points to look for are "well-bred" heads-i.e., showing an approach to the type of some improved beef breed-wide backs, thick flesh, and good middles.

Turning to Milk Production, it is of course true that large cows, on the average, yield more milk than small ones. Moreover, it seems to be true that, on the average, large cows yield more milk in proportion to their total food consumption than small ones. The next point may be illustrated by an interesting analysis that was made by Eckles[1] of the differences between a good and a bad Jersey cow. He found that there was very little qualitative difference in digestive capacity i.e., the two animals extracted practically the same quantities of nutrients from like quantities of particular foods. The maintenance requirement was not quite the same, but in the case in question the poorer cow had the advantage *i.e.* it maintained its weight when not milking on rather less food than the better animal. In the investigation the cattle were fed for a year in such a manner that their body weights remained practically constant. The important conclusion was that the milk yield was proportional, not to the total ration but to the amount of food eaten over and above the amount required for maintenance; and hence, the heavier yielder was also the more economical producer.

If we look for the causes of unsatisfactory yields it is obviously possible that these may be of three kinds. Firstly, certain cattle, milk well for a short time, but "milk down" in condition very rapidly and soon drop

[1] See Eckles's Dairy Cattle and Milk Production.

off in yield. Here it seems that the mammary capacity may be ample, but the digestive organs are not able to work at sufficiently high pressure to supply the necessary raw materials. Secondly, there is the type represented in many cows of good beef form. They have ample feeding capacity but the surplus food is diverted to body fat instead of to milk. If a heavy ration be fed to such an animal it continues to fatten upto a certain point, when the appetite declines, the maintenance energy requirement rises, and a balance between the income and expenditure of energy is restored. In this type it may simply be that the mammary system is not of sufficient capacity to deal with the available nutrients, and that, failing an outlet as milk, the surplus nutrients are converted into body fat. This, on the whole, appears likely. On the other hand it is possible that there is a kind of competition for the available nutrients between the fat-storing cells of the body and the mammary cells, and that in animals that have a marked tendency to fatten, the process of storage tends to take precedence over that of milk production. In this last case we should have to regard the fattening tendency as definitely opposed to high milking capacity. Unfortunately there is no evidence on which to reach a definite conclusion. Some deep-milking cattle do, however, fatten readily when dry.

Thirdly, the activity of the mammary tissues depends on the stimulus of parturition, and a new cycle begins with each calving. The yield during the lactation may be determined, to a considerable extent, by the amount of response to this stimulus, which is caused by internal secretions. Hence, these secretions may have an important bearing on milking capacity. It has, in fact, been shown that milk production can be stimulated-even in maiden heifers-by the injection of hormones, particularly synthetic substances, such as hexoestrol and stilboestrol. Experiments have shown that it is possible, by feeding thyroxin, to increase milk yields by several pounds per day-over a period of several weeks; but the effect on the total lactation yield is very small.

The importance attaching to the compositional quality of milk depends on the mode of utilization. Where whole milk is sold, all that is ordinarily necessary is that the milk should conform (with a reasonable margin of safely) to the British presumptive standard of 3 per cent, butter-fat and 8½ per cent, of other solids. Occasionally a market, at a higher price, can be found for milk of superior quality; under the existing milk-marketing scheme special prices are paid for the milk of Channel Island and South Devon cattle. The consumer also prefers milk with a deep creamy colour, which is due to its content of carotene. The colour of milk is partly a matter of the breed and individuality of the cow, but is also influenced by the carotene-content of the food. The yield of butter or of cream depends mainly, and the cheese yield depends very largely, on butter-fat content; so that for these purposes the composition of the milk is important, and attention must be given to the matter in breeding. There is a slight negative correlation between butter-fat content and yield i.e. cattle tot give rich milk are generally somewhat smaller producers than those that give poor.

It will obviously be difficult to translate into terms of bodily points the physiological qualities that go to make a good cow, and, as a matter of fact, the judging of milk cattle from their appearance is at best rather an unsatisfactory business. The real test is actual performance, with due allowance for age and the other conditions that are known to affect yield. The mean yield over a series of lactations and under ordinary farm conditions is the best kind of record. Failing this, we must take the record over a shorter period and endeavour to form an impression, from the cow's appearance, of her constitution and general ability to live and thrive in the average environment of her breed.

Of the outward signs of milking capacity the most reliable are the size and "quality" of the udder. The udder should be large, extending well forward along the belly and well backwards and upwards. It should also be wide and deep. A pendulous udder, if short and narrow, is often far less capacious than it appears. The udder should be composed largely of actual secreting tissue, which is elastic, so that the udder is large before milking and has a shrunken appearance and a soft, pliable feel when empty. The teats should be placed well apart and be of a convenient size for grasping i.e. say, from 3 to 4 in. long. After the udder, the size of the milk veins, which extend forward from the udder along the belly, is probably the surest guide among the external characters; but it must be remembered that the milk veins become progressively larger with each lactation A capacious middle, as showing ability to feed, is also a point of some importance, and a quiet, docile temperament, as indicated by the head, eye, and general behaviour, is desirable.

While no substantial correlations have been established between milk yield and individual body measurements, it has been shown tot there is a fairly high correlation between performance and the total "score" for conformation awarded by a competent judge.

The conventional "ideal dairy cow" has many other points that are considered to be indications of deep-milking qualities, but the value of many of these is doubtful, and there is no conclusive evidence in favour of any of them Many cows of "dairy type" are good milkers, but others that are poor, as judged by the accepted standards, milk well nevertheless. Briefly, the " ideal dairy cow" is wedge-shaped-i.e. she is narrow in the fore end, with sharp withers, and wide in the hind quarters; also, she is relatively shallow in front and deep behind. The horn should be fine and have a waxy appearance, the eye large and placid, the head broad between the eyes-rather longer than in the beef type-and lean; the neck should be long, thin, and muscular, the shoulder light and sloping, the withers sharp and narrow; the chest must be of fair depth and not too narrow, otherwise the animal is likely to be lacking in constitution; the middle should be large, the ribs deep, well sprung, and placed wide apart with a wide space between the last rib and the hook; the hind quarters should be lengthy and wide, but the hooks not specially prominent; the thighs should be thin and the tail long and fine. Generally, the animal should be thin in flesh, having a somewhat angular but refined appearance; the skin should be thin, soft, and elastic. The colour of the milk is correlated to some extent with that of the skin secretions, as seen, for example, inside the ears and on the end of the tail. Animals that have deep yellow skin generally give deep-coloured, rich looking milk. Sound legs are necessary if the animal is to have a long milking life. The hind legs should be broad above the hock, and the lower part of the leg should be perpendicular when the cow is standing naturally. The hocks should be wide apart and the gait should be free and easy.

Dairy type is somewhat differently interpreted in the several dairy breeds. The most extreme type is suited only to the best conditions. It is recognised that cattle which are expected to withstand harsh weather or survive periods of scarcity must be rather less refined in form and carry more flesh than the "ideal" dairy cow. Indeed, there is no more important principle in livestock improvement than that of adapting the type of animal to the conditions under which it has to live. The very productive strains of dairy cattle, like the early-maturing strains of beef cattle, demand a very high level of nutrition. If this is not provided the cow's high productive capacity becomes a danger to herself.

Unreliable as some of the "signs of milk" are when applied to the cow in milk, they are of still more doubtful value as a measure of the hereditary milking qualities in the bull. The most reliable measure of a bull's worth is the performance of his female relatives, whether dam, sisters, or-most important of all-daughters.

The dual-purpose type represents an endeavour to combine milk production in the breeding cow with good fattening qualities in the males and non-milking females of the breed. The recent development of dual-purpose types-e.g. in the Shorthorn-has been regarded by some as a retrograde step: they regard specialization as essential to the progressive improvement of livestock and look upon the dual-purpose animal as at best an unsatisfactory compromise between two essentially opposite types. But mere is no proof of actual incompatibility between deep-milking qualities and the tendency to fatten when not in milk, and it is, in fact, possible to combine to a considerable extent been qualities and milking tendencies in the same animal. In any case it would seem that dual-purpose cattle are an economic necessity, for under intensive conditions and with the increasing size of the dairy industry there can be no cheap or abundant supply of store cattle except those bred from dairy cows. The efforts that are being made to fix and perpetuate the combination of qualities are meeting with a large measure of success. A breed, the cows of which give an average yield of the order of 8000 Lb. while the steers yield good carcasses at under two years old, would seem to be possible.

BRITISH BREEDS OF CATTLE

There are, all told, twenty-two British breeds of cattle, each of which has its own herd book. No classification can be altogether satisfactory, because each breed shows a certain amount of variation in type from herd to herd, according to the purpose for which the herd is kept, the method of management, and the fancy of the particular breeder. Thus, there have been dairy herds of Herefords and beef herds of Friesians. We may, however, fairly accurately classify the Hereford, Aberdeen-Angus, Sussex, Galloway, and West Highland as beef breeds. The Devon may also be included, but a fair proportion of the cows yield more milk than is required for the normal development of the calf. There are six dairy breeds, the Friesian, Ayrshire, Jersey, Guernsey, Kerry, and Gloucestershire, the last being numerically unimportant. The major dual-purpose breeds are the Red Poll and South Devon, and minor ones are the Long-horn, Belted Galloway, Dexter, and Park. The Shorthorn consists of two main types-the one, represented mainly by animals of Scotch blood, is of extreme beef form, while the Dairy Shorthorn is really a dual-purpose animal. The position is similar in the Lincoln Red Shorthorn and the Welsh Black- *i.e.* each contains beef and dual-purpose strains.

The Shorthorn (Plate 1 A) originated in a number of closely related types that existed during the early eighteenth century in Yorkshire, Northumberland, and Durham. The centre of most of the early work of improvement was the area within a radius of about twenty miles of the town of Darlington. The ancestry of these foundation types, while not fully known, appears to have been mixed. The blood of the old black Celtic, of the red Anglo-Saxon, and of the broken-coloured Dutch cattle seems to have been blended in them. There are definite records of importations from Holland as late as 1750, and it is certain that there were many others of earlier date. The cattle of Tees water, a century and a half ago, were held in high esteem both for beef and milk, perhaps more especially the latter. They were of large size, short horned, and of mixed colours, with red-and-white pre-dominating. The Holderness breed was described in 1743 as haying "wide bags, short horns, and large bodies, which render em-whether black or red) the most profitable, beasts for the dairy-man, grazier, and butcher."

Some careful selective breeding had been done before 1750, but the earliest of the great improvers were the brothers Charles and Robert Colling of Ketton and Barmpton, near Darlington, who began breeding about 1780. They selected the best available animals of the Teeswater type, devoting special attention to fattening qualities, and they inbred very closely. At Charles Ceiling's famous sale in 1810 the very much inbred bull "Comet" realized 1000 guineas.

The Booths of Killerby and Warlaby in Yorkshire were also, for three generations, famous improvers. They used Colling bulls in the early stages, and produced a type of animal notable for its substance, wide back and thick flesh, but of little account as a milker. Booth cattle were very prominent in the showyard, and were highly prized, over the long period from 1840 to 1880.

Thomas Bates of Kirklevington commenced breeding in 1800, and originated a type of Shorthorn that rivalled the Booth in popularity. His ideal was a dual-purpose animal, and he attached great importance to handling qualities, to style and character, and to particular blood lines. He inbred very intensively Twenty years after Bates's death, which occurred in 1849, his cattle enjoyed a great boom and brought extraordinary prices. After 1890 they fell from favour, but now again Bates's blood is in high esteem among breeders of the dairy type.

The Scotch Shorthorn is mainly the product of the efforts of Amos Cruickshank (1808-95) of Sittyton, Aberdeenshire. He was no partisan of the Boom or of the Bates type, and founded his herd by selecting good animals of very varied breeding, Booth blood, on the whole, predominating. His special aims were a robust constitution, good middle, thick flesh, early maturity, and feeding qualities generally. The Scotch type rose very rapidly in favour-first in the United States and, after 1890, in Britain. It remains the popular type of Shorthorn in beef-exporting countries like the Argentine, and British herds are maintained partly with the object of supply ing the demand for bills to export to these countries.

The last outstanding step in Shorthorn history was made in 1905, when a group of breeders started the Dairy Shorthorn Association, with the object of reviving the original dual-purpose type which at that time seemed to be in danger of extinction owing to the high prices prevailing for animals of the Scotch strain. The Dairy section of the breed has made rapid progress since that date.

All British Shorthorns are registered, without regard to type, in Coates's Herd Book, established in 1822. This is an "open" register- *Le,* animals are eligible for entry if they have a certain number of registered crosses, and need not be descended from registered parents on both sides. The regulations require, for females, four crosses of registered blood, and for males five, so that it is possible by the use of registered bulls for tour or five generations to grade up non-pedigree herds until all their members are eligible for entry. Any registered animal may compete in the classes for "Shorthorns" at the leading shows, but males can be shown in the "Dairy" classes only if their dams have passed certain specified milking tests. Cows in milk are also required to produce certain minimum yields.

The special breed points applicable to both types are as follows Colour: red, or red-and-white, or white, or roan. Deep cherry red and deep uniform roans arc preferred; patchy roans, red-and-white, and yellowish or gingery red and roan are regarded as objectionable. The muzzle should be flesh-coloured a black or cloudy nose being very undesirable. The horn is short, laterally flattened, of fine waxy texture, and free from black. It should not curve upwards, but only forward in the male and forward and inward in the female, remaining about level with the crown.

The *Beef Type Shorthorn* (which, in fact, practically means the Scotch strain) is a large animal, exceeded in size only by the South Devon Compared with that of most other beef breeds the conformation is more nearly rectangular, the back being wide and very flat and the buttock very square in cut. The type shows remarkable early maturity, but the carcass tends to be '"wasty" owing to excess of external fat. The cows are in general poor milkers, sometimes yielding insufficient milk for the normal nourishment of their calves. The coat is abundant and the cattle are reasonably hardy.

The great reputation of the beef shorthorn rest on its wide adaptability and on its capactiy to stamp its characteristic upon in-ferior types of animal. It has been the universal experience that where the breeder has had of necessity to begin with thinfleshed, late-maturing "scrub" cows as a foundation, "one cross of the Shorthorn has been equal to two of any other breed."

It has been widely used for this purpose in building up the export beef trade of countries like Argentina.

The *Dairy Shorthorn* is numerically very important, but in recent years has been losing ground to the specialized dairy breeds. Mature cows in breeding condition weigh about 1200 lb. The general build, as illustrated in Plate 1.B, is deep-bodied and roomy, with long, level hind quarters, moderately fine shouders, and broad back. The thighs are more fleshy than in the purely dairy breeds. The udder is capacious, and the present tendency in the show yard is to attach great importance to a square level bag with neatly placed and moderately small teats.

The better herds of Dairy Shorthorns, with good feeding, reach average yields of the order of 7000 to 9000 Lb. per annum; individual records of over 10,000 Lb. are common, but those over 20,000 lb. are rare. The butter-fat content of the milk probably averages about 3.6 per cent, but in some districts it has deteriorated in recent years. This has been due to the tendency to select for heavy yields without regard to the composition of the milk in such cases it would be easy to attain higher quality by selective breeding.

An Advanced Register for Dairy Shorthorn cows was set up in 1947. The milk yields required for admission are :

1. After calving at four years of age or more, at least 10,000 lb. milk and 375 lb. of butter-fat
2. After calving at three to four years of age, 8500 lb. milk and 330 lb. butter-fat.
3. After calving at less than three years of age, 7500 lb. milk and 300 lb. butter-fat.

Higher yields are required where thrice-daily milking is carried out, and the quantities specified must be produced in not more than 315 days. Cattle accepted for the register must be of good general breed type, and special regard is paid to udder conformation.

Dairy Shorthorn steers, when adequately fed, make good quality beef at about two years old, and selected animals can be finished at eighteen months. When kept as "stores" until two or two-and-a-half years old, and then fattened, they tend to be too large for the average family-butcher's customers, and are rather rough and bony in conformation.

The Lincolnshire Red Shorthorn, or Lincoln Red, is a strain of Shorthorn found chiefly in its name-county. It is of a whole cherry-red colour, large in frame, and of robust constitution. The Lincoln Red, like the Shorthorn, consists of beef and dual purpose strains, which are only rarely interbred. Separate classes are provided at the leading shows for cows of the dual-purpose type. The Lincoln Reds are the prevailing breed in many parts of Lincolnshire, Nottinghamshire, and the adjoining parts of other counties.

PLATE 1

A. SHORTHORN BULL

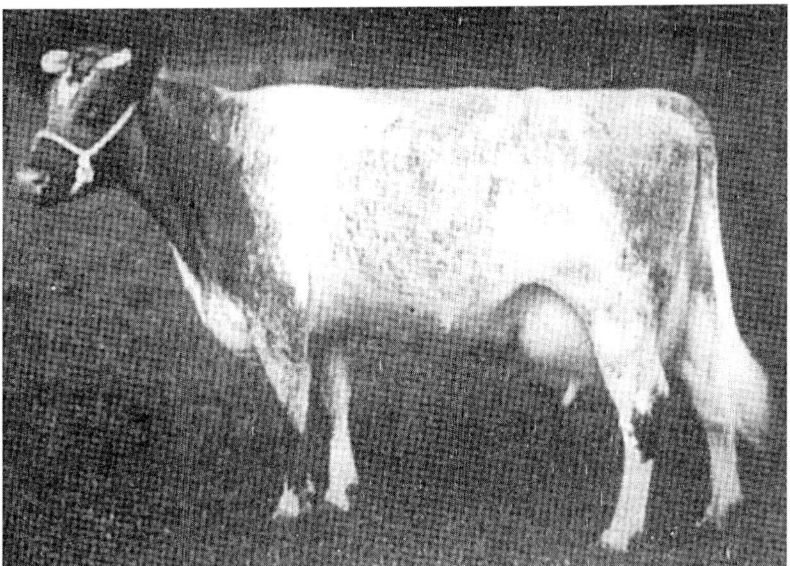

B. DAIRY SHORTHORN COW

PLATE 2

A. ABERDEEN-ANGUS BULL

B. HERFORD BULL

PLATE 3

A. SUSSEX BULL

B. DEVON HEIFER

PLATE 4

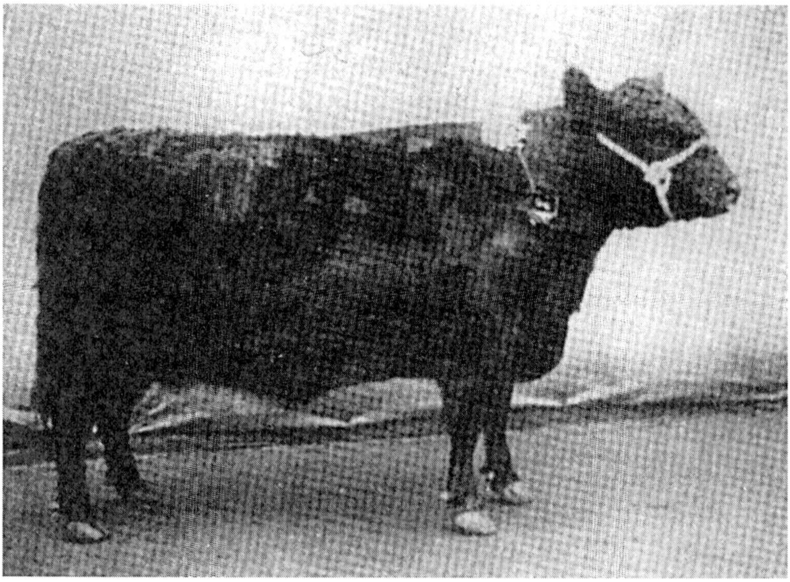

A. GALLOWAY HEIFER

B. WEST HIGHLAND HEIFER

The Northern Dairy Shorthorn is a recently formed type for which a herd book has lately been established. Its home is in the Dales of Yorkshire and Durham and the upland parts of Lancashire. It is smaller, more lightly built and more active than the typical Dairy Shorthorn and is said to be better adapted to the hilly land and the rather severe climate. Ayrshire blood has been used in the formation of some of the herds.

The Aberdeen-Angus (Plate 2, A) originated in the old local breeds of Angus and Aberdeenshire. As far back as records go, which is to the middle of the sixteenth century, a proportion of these cattle have been black and hornless, characteristics tot have now been practically fixed in the breed. Red calves, and others with white markings, are dropped occasionally, but horns have been completely bred out. The earliest improver was Hugh Watson of Keillor, near Coupar-Angus, who laid his foundation in 1808. Of the other breeders in both counties who carried on the work the greatest was William M.Combie (1805-80) of Tillyfour in Aberdeenshire, who not only improved the breed but succeeded in making its merits generally known. The chief breeding district is still the country lying between Inverness and Perth and eastwards to the sea, the breed being restricted, in the more mountainous parts of this region, to the straths and glens. Speyside has very numerous herds. There are now many Angus herds throughout the arable districts of Scotland and also in most counties of England and Ireland. The herd book, established in 1862, is a closed register i.e. no animals may be entered except, such as are the progeny of parents already registered.

The Aberdeen-Angus is black in colour, with a brownish tinge in the winter coat and in that of the calf, and with often a little white on the underline behind the navel. It is hornless, with a somewhat sharply pointed poll. The Shorthorn, Hereford, and Sussex all exceed it in size, but the difference in weight is less than it appears, as the Angus is compact and heavy for its size. The Smithfield weights show that the breed is actually about 4 per cent, lighter than the Shorthorn. The bone is small, and the general symmetry and smoothness of outline are marked characteristics. The back and also the buttock are somewhat more rounded than in the Shorthorn, and the hooks scarcely project from the general line of the side.

As a commercial animal the Angus is an early-maturing beef beast of superlative quality. The dressing percentage is high, the carcass is very free from patchiness, and the beef is fine in grain and well marbled. In open competition at the larger fat stock shows during the past thirty years the breed and its crosses have won more championships than all others together. The yield of milk, though irregular, is high for a beef breed, and cows with really good dairy qualities occur. The temperament is rather more active than that of most beef breeds.

The Angus is particularly well adapted for intensive feeding and for "baby beef production. It has neither proved so widely adaptable as the Shorthorn, nor so useful under poor range conditions as the Hereford. For grading up thin-fleshed and slow-maturing types it is less useful than the Shorthorn, but for imparting quality to herds that are already improved to some extent, it is excellent. Angus bulls are largely used for crossing with cows of dual-purpose type in the production of commercial beef cattle. If still higher quality is required, the cross-bred females may be mated to pure-bred Angus bulls, producing three-quarter-bred progeny. The breed has been an outstanding success under arable or semi-arable conditions-for example, in the maize-growing districts of the United States. It has been exported in large numbers. Mated with a white Shorthorn bull the Angus cow produces the well-known and highly valued blue-grey crossbred.

The Hereford (Plate 2, B) - As early as the beginning of the seventeenth century the cattle of Herefordshire were held in high esteem both as work oxen and for beef. They were large, stoutly built, hardy, and generally well suited to be reared and fattened on the fine pastures of the county. Improvement in the direction of early maturity and duality of meat began early. The names of Benjamin Tomkins the younger (1745-1815) and of John Hewer (1787-1873) are the best known-the one amongst the older, the other in a more recent generation of breeders. The herd book was established in 1846 and has been closed since 1883.

As regards Britain, the Hereford is rather a local breed. Most of the prominent pedigree herds are found in the English and Welsh counties adjoining the border, from Cheshire in the north to Monmouth and Gloucestershire. The breed is, however, largely represented in Ireland, and has a large place in the United States, Canada, South America, and

Australia. The breed characteristics are a large size (little inferior to that of the beef Shorthorn), a deep rich red colour (without any black or dark brown hairs), with white head and underline, and generally a narrow band extending along the upper part of the neck and over the crops. Some cattle have a red spot at the eye, which is an advantage in countries of glaring sun. The horn is of medium length, slightly drooping and of a waxy appearance to the tip; the back is wide and thickly covered, the chest deep and capacious, and the legs short. The skin is thick but mellow, the coat mossy and very dense and abundant. The Hereford has neither the superlative quality of meat of the Angus, nor is it distinguished, even among the beef breeds, for milking qualities. Its chief value rests in the fact that it has, of all breeds, the fullest combination of early maturing and fattening qualities with a robust constitution. It is an excellent "doer" under pastoral conditions, will fatten readily on good grass alone, and abroad has been found to live and thrive on the poorer and drier type of range where no other improved breed can exist. In their home country the cattle are very generally kept out of doors throughout the year, and as a consequence the breed has always been relatively free from tuberculosis.

The Sussex (Plate 3, A) is native to the Weald district and the marsh lands of Kent, Surrey, and Sussex. The breed was originally kept mainly for draft purposes, and continued to supply work oxen for the Wealden clays long after the horse had become the sole draft animal elsewhere. With changing conditions the Sussex has become a beef breed, but it still bears some traces of its centuries-long use in the plough. In weight it is practically equal to the Hereford, but it is less compact, somewhat more muscular in build, and rather later maturing. The colour is a deep red, the horn fairly large, spreading, light at the base and dark-tipped; the skin is fine and the coat rather short.

The general conformation is that of a good beef animal, the quality of the beef is excellent, and in hardiness and ability to thrive on poor fare the breed is outstanding. It has met with considerable success in South Africa. The herd book was established in 1874.

The Devon (Plate 3, B) is native to the hilly region in the north of Devonshire, but spread at a very early date into the adjoining parts of Somerset. Many descriptions dating from the latter part of the eighteenth and early nineteenth centuries show that the breed existed in those days in a form not unlike the modern type. The Devons were described by Marshall (1796) as the best workers he had ever seen, and by John Lawrence, in 1805, as having "for a century past commanded the best price at Smithfield." All writers agree that they were hardy, excellent draft animals, and good feeders. Francis Quartly (1764-1856) of Great Champson, Molland, was the greatest individual improver. During the period between 1793 and 1823, when war prices tempted many of the breeders to sell their best cattle and to slaughter their calves for veal, he gradually got together a herd of the best individuals he could find. By continued selection and inbreeding he produced a strain that has since been regarded as the fountain-head of all the best blood.

At present time the Devon is most numerously represented in North Devon and Somerset, though herds are scattered fairly widely over southern England. On the lower and more fertile pastures of Somerset the cattle are larger and of greater substance, but lack to some extent the refinement of the upland cattle. In appearance the Devon rather closely resembles the Sussex, but is considerably smaller, of a rather brighter shade of red, very frequently with dapple markings. The horns are of moderate length, with an upward tilt in the female. They are of a brownish shade with darker tips. General symmetry of build and smoothness of flesh are marked characteristics. The breed is hardy, well adapted for grazing, and produces beef of choice quality. In recent years more attention has been devoted to dairy qualities, and several pedigree herds are now kept for dairy purposes. In the ordinary showyard, however, Devons are still judged as beef cattle. The breed has met with some success abroad, particularly in semi-tropical countries.

The Galloway (Plate 4, A) is an old breed of obscure origin. Two hundred years ago it occupied, to the almost complete exclusion of other breeds, a very largle area in the south-west of Scotland. The extension of dairying in this district has caused it to be largely displaced by the Ayrshire, but on the other hand it has extended into the north of England. Like the Aberdeen Angus, the Galloway is black, polled, and of beef type; otherwise it has few points of similarity with the northern breed. In the Galloway the head is very short, with a broad poll, and with the ears set on rather low; the neck is rather long; the body is deep, but longer and narrower than in the most highly developed beef breeds. A certain slackness of back and prominence of shoulder are common faults among the ordinary commercial individuals. The skin is thick and the coat very abundant, consisting of a very dense mossy undercoat with an outer covering of long fine hair.

The Galloway is one of the hardiest of breeds, whether under the cold and very wet conditions that obtain in winter among its native hills or in districts of intense winter frosts. The cows of most herds live out of doors throughout the year, being brought to lower ground in winter and returning to the higher altitudes soon after calving. The quality of the meat is excellent, and the carcass is neither patchy nor wasteful from excess of fat. In recent years the breed has achieved great success in the carcass competitions at Smithfield Show. The Galloway is, naturally, somewhat slow to fatten, yet wonderful examples of early maturity are often seen at fat-stock shows. Judged by the Smithfield weights, the Galloway is nearly 15 per cent, lighter than the Shorthorn at the same age. The Polled Herd Book was originally founded as a common register for Aberdeen-Angus and Galloway herds, but a separate Galloway Herd Book was started in 1877. A dun-coloured strain of the breed is registered in the same herd book.

Crossed with the white Shorthorn the Galloway produces a "blue grey" that is hardier, if somewhat slower in maturing, than the corresponding Angus cross, and equally esteemed for the quality of its beef. The Galloway is not noted for dairy qualities, though deep-milking' cows do occur.

The Belted Galloway is a distinct strain of the breed, separately registered. It has a characteristic white belt round the middle of the body. Some herds are bred upon beef and others upon dual-purpose lines. Dun animals carrying the white belt also occur.

The West Highland (Plate 4, B), or Kyloe, is the native breed of the Western Highlands and Islands of Scotland. The recognized colours embrace various shades, of red, yellow, dun, and cream, as well as brindled and black, which last formerly prevailed but is now rare. The breed is of moderate size, with large spreading horns and a shaggy coat of straight or slightly wavy hair, and a dense undercoat of a fine woolly texture; the form is square and deep, but somewhat lacking in width. The West Highland is exceedingly hardy and well adapted to live out of doors in the wild and exposed surroundings of its home country, where it can subsist on the poorest of grazing. It is a slow grower and difficult to fatten until it is fairly mature; indeed the steers are rarely marketed fat until they are three-and-a-half years of age. The beef of well-fattened beasts is of choice quality. The milk yield is low, being little more than sufficient for the rearing of the calf.

DUAL-PURPOSE BREEDS

The Red Poll (Plate 5, A) of East Anglia was produced by the blending of two pre-existing local types, viz. the Norfolk Horned, a small and compactly built red beef breed, and the Polled Suffolk Dun, which was much praised by early writers, including-Arthur Young, for its exceptional dairy qualities. The blending was carried out on a systematic plan and with the avowed object of producing a "new sort" of dual-purpose cattle. The work began in the first decade of last century, when two breeders, John Reeves and Richard England, both of Norfolk, began on the lines indicated. About the same time a breeder named George, farming in the same county, founded a herd of red individuals selected from among the Suffolk Polled cattle. The breed was first recognised by the R.A.S.E. in 1862, and an admirably conducted herd book was started twelve years later. The Red Poll is still far commoner in Norfolk and Suffolk than elsewhere, but there are herds in almost every part of England and a few in Scotland.

The colour is a uniform red, with frequently some white on the udder of the female, and a white switch. In size the Red Poll equals the Angus, though it carries rather less flesh and is, therefore, lighter. Horns or "scurs" are not permitted, and a dark nose is regarded as very objectionable; both of these departures from type are rare. Generally the Red Poll is a well-built and symmetrical dual-purpose animal, although the udder is often not so shapely as could be wished. In the past there has been some irregularity of type, due to the varying importance attached by individual breeders to beef conformation and milking qualities respectively, but great progress in the direction of uniformity has now been made. The steers fatten readily at an early age and produce a good quality of beef In yield and quality of milk many herds compare favourably with those of dairy breeds. Good mature cows will give 8000 lb., and many high individual records, up to twice this amount and more, have been made.

As a "farmer's cow," particularly on moderately poor land, the Red Poll is a valuable animal. The breed is known in the United States and in South America, while there is a growing demand from many of the Dominions, especially the Union of South Africa.

The South Devon (Plate 5, B) is numerously represented only in Devon and Cornwall. It differs very markedly from the Devon in being of much greater size. The original colour was a pale brownish red, but the preference is now for the deeper shades. The South Devon is the largest of British breeds, cows in ordinary breeding condition frequently scaling 14 cwt., and bulls in show condition having been known to exceed 30 cwt. The general type, according to the ordinary standard of cattle judges, is rather coarse. Improvement is of comparatively recent date, and a herd book was instituted only in 1891. Milk yields of over 1000 gals, are quite common-several lactation yields of over 2000 have been recorded-and cows of the breed have done well in open milking trials. The average butter-fat content is about 4.3 per cent, and the colour of the milk is a rich creamy yellow. The breed has been largely exported to South Africa, where it is popular on account of its suitability to the climatic conditions and its usefulness for producing trek oxen

The Longhorn was, in Bakewell's time, very widespread over the west midland and western counties of England, from Westmorland to Somerset and from Shropshire to Leicestershire. The best types *were* found in Lancashire and in the Craven district of Yorkshire. At the same period the breed was very widely distributed in Ireland, whither it had probably been introduced from England. Before it passed through Bakewell's process of improvement the Longhorn was a large and powerfully built animal that produced good draft oxen and fair milk cows. It had a rough shoulder, flat sides, and large bone, was difficult to fatten when young, and produced a dark-coloured and coarse carcass. At the sacrifice of its dairy qualities Bakewell made it an early maturing beef breed of greatly improved quality, though even in Bakewell's beasts the fat was not well laid on and the meat was poorly "marbled." In the middle part of last century the breed was all but driven out of existence by the Shorthorn. The numbers then increased slowly for a time but are again very small. The breed is now found principally in Warwickshire and the adjoining counties. On the poorer sorts of clay land it has proved a useful dual-purpose animal. Under these conditions it is hardier than the Shorthorn, and gives a creditable yield of milk of good quality. The breed is of medium size, still somewhat rough in conformation, with long and generally drooping horns. The favourite colour is a dark red brindle, and the characteristic markings include a white stripe along the back, a good deal of white on the underline, and a white spot on the thigh.

The Welsh Black (Plate 6, A) The black, horned cattle of Wales are of great antiquity, although there is evidence that cattle of other colours have existed in the Principality, along with them, from medieval times. Formerly two breeds were recognised. The North Wales or Anglesey was a hardy, slowmaturing beef breed, well known during the past two centuries as producing the highly valued grazing cattle imported into the Midlands for fattening, under the name of "Welsh Runts." In the south was the Pembroke or Castlemartin breed, rather longer and looser of frame and thinner of flesh, prized chiefly for the dairy. The herd books were amalgamated in 1904, when the breed name of Welsh Black was adopted. Of medium size (about similar in this respect to the Red Poll), the Welsh carries rather long, spreading horns. Show specimens often leave little to be desired in respect of symmetry, but the ordinary run tend to be rather high at the rump, rather weak in the thighs, and hard " handlers." The steers are generally marketed fat at three years old, when they produce a carcass that is full of lean meat and of very high quality. As said earlier, there are varying degrees of emphasis on beef and mille. Some herds are kept like Galloways, the cows suckling their own calves only. Some are crossed, commonly with a Hereford bull, in order to obtain a heavier and earlier-maturing type of store. Other herds, kept on marginal rather than hill land, are kept mainly for milk production.

The Blue Albion has its home in the High Peak district of Derbyshire. It seems to have originated mainly from an admixture of Shorthorn and Welsh blood, but a good deal of Friesian influence can be traced in many herds. The general type approaches that of the Dairy Shorthorn, but the recognized colour (which, however, does not breed true) is blue or blue-roan with more or less white. The breed has something of the hardiness of the Welsh, and many of the cows are deep milkers. A Breed Society was formed in 1919.

Park Cattle : The Park Cattle or "Wild White Cattle," whose origin has been the subject of much discussion, are now under a Breed Society, have a herd book and have obtained recognition from the Royal Agricultural Society. The cattle are white with black or red "points." Some herds are polled and others horned. The general type is dual-purpose & the breed has several records of over 1000 gals, to its credit.

DAIRY BREEDS

The British Friesian (Plate 6, B) is a branch of an ancient breed that has its home in the Dutch province of Friesland, lying north-east of the Yssel Lake (the Zuider Zee). It is only one of several Dutch breeds, but is probably the most highly improved, and certainly possesses the most pronounced dairy qualities. Dutch cattle of the Friesian type have exerted a large influence on other breeds, including our own Shorthorn and Ayrshire, as well as a great number of the lowland breeds of Western Europe. At the present time the Friesian is second only to the Shorthorn in its wide and general distribution throughout the world. The home province consists largely of low-lying and fertile pastures, and the special centre of the breed, the district round Leeuwarden, is one of the exceptional fertility, though its climate is distinctly raw and cold.

During the second half of last century large numbers of blackand-white Dutch cattle, both Friesians and animals of the closely related North Holland type, reached Britain. The cows were chiefly bought for town dairies, and only a small proportion were in consequence kept for breeding; but some herds were established in the early seventies and again about 1890. Few of these were kept strictly pure or managed with any particular care, and the breed, as a whole, made little progress until the Breed Society was established in 1909. Thereafter it advanced in favour with extraordinary rapidity, and has continued to increase in numbers ever since. The foundation stock was accepted for the herd book by inspection, and it was impossible to prevent the admission of a proportion of animals of doubtful ancestry. This initial difficulty has since been largely overcome by very strict insistence on the true Friesian markings and type, and by the importation from Holland in 1914, 1936, and 1950, from South Africa in 1922, and from Canada in 1946 of numbers of specially selected animals.

The Friesian is by much the largest of the dairy breeds. The standard colour is black-and-white in clear and distinct patches, though an occasional dun-and-white animal may be seen. A separate herd book has been set up for the red-and-white variety. The feet and switch must be white, and a white star or blaze on the forehead is popular and of frequent occurrence; the head is rather long, the horse short and small, with a forward and inward curve; the neck is rather shorter than in the other dairy breeds; the hind quarters are exceptionally wide, and the animals when young or when out of milk carry more flesh on the back and thighs than do most dairy cattle. The mammary system shows great development and the milk veins are very prominent, but the udder is less square and regular than in the Ayrshire. The breed has sometimes been regarded as rough in conformation and lacking in quality, but these defects are being largely eliminated.

The special merit of the Friesian lies in its great milking capacity. In the United States and elsewhere the breed holds most of the records for quantity, but the record British yield up till 1954 (41,644 lb. in 365 days) was held by a Dairy Shorthorn. In this country many individual yields of from 2000 to over 300 gals, have been recorded, and large herds have averaged over 1000 for cows and heifers together. One cow of the breed has yielded over 3000 gals, in each of three successive lactations, The breed has provided nine-tenths of the total number of official records of 2000 gals, or better in a year. The fat-content of the milk is commonly lower than with other breeds. In Holland, the average was at one-time as low as 3.1 or 3.2 per cent. In Britain, while occasional animals fail to reach the 3 per cent, standard, the average would be more nearly 3.4. There is, of course, much individual variation, and a little careful selection will usually succeed in bringing a herd average upto the normal 3.7 or 3.8. Leading breeders are now devoting close attention to the quality of the milk.

As a meat producer the Friesian is the best of the purely dairy breeds, if it be considered as such; but it is not so satisfactory as, say, the recognized type of Dairy Shorthorn. The calves are large and feed rapidly into excellent veal. The steers, unless they are well fed from birth, tend to become too large before they can be fattened out. But if given a good start, they provide very useful carcasses.

The Friesian is quite ordinarily robust and hardy, but is pre-eminently suited to the better sort of pastures or to arable dairy farms where food is abundant. On poor upland pastures it has been found that it cannot successfully replace the Ayrshire.

The Ayrshire (Plate 7, A) came into existence some time before the end of the eighteenth century, but of the methods by which it was evolved we have no clear idea. The native cattle before the new improved breed displaced them were miserable beasts, "few of the cows yielding more than from ½ to 2 gals, in the day, at the height of the season, or weighing when fat more than 20 st. Teeswater i.e. early Shorthorn-cattle were imported, and possibly other cattle of more immediate Dutch extraction. There are traditions, too, of "Alderney" i.e. Channel Islands-cattle having been introduced. The centre of improvement was the parish of Dunlop, and John Dunlop of Dunlop was one, and probably the greatest, of the early breeders. The breed, as it spread over the northern portion and finally over the whole of the county, was successively known as the Dunlop, the Cunningham, and the Ayrshire. It began to be generally recognized as a distinct and improved breed about 1800, and has ever since been famous as a milker. Thus, among the less extravagant estimates of the average yield is that of Sir John Sinclair, who, in 1813, reckoned it at 900 gals., and that of Low, in 1845, who states that "healthy (mature) cows on good pasture" should give between 800 and 900. In the beginning of the present century some strains of Ayrshires deteriorated owing to the neglect of dairy qualities for purely "show" points. On the other hand, the milk-recording scheme initiated by John Speir, in 8½ has had a very beneficial effect on the utility qualities of the breed.

The Ayrshire is a comparatively small breed, cows in ordinary breeding condition scaling between 8½ and 9 cwt, or, say, four-fifths of the average Friesian weight. Cattle on poor upland farms are, of course, lighter than those found in the more fertile districts. The colour is red or brown with a varying amount of white a large proportion of white being fashionable. Black-and white, which is not uncommon among commercial cattle, is an accepted colour, but is not in favour with pedigree breeders. The horns are rather large, set fairly wide apart; and have a very characteristic upward curve. The conformation is very symmetrical and neat, the back being straight and the lines of the body smooth. The udder is also characteristic, being of great length and width, firmly attached (not pendulous), level in the sole, with small teats squarely and neatly placed. At one time the fashion for 'tight vessels" and small teats was carried too far, but the modern show standard is quite compatible with high production.

On the poorer sorts of land, at high elevations and in cold climates, the Ayrshire is a very valuable dairy animal. The general average of all the yields recorded in recent years is over 650 gals, with about 3.8 per cent, of butter-fat. Many herds (whole herds, including frequently 20 or 25 per cent, of heifers) have averaged over 800 gals, with a butter-fat content of 4 per cent, or better. Individual yields of 1000 gals, and over are common, so tot, in view of its small size, the Ayrshire must be regarded as a highly economical producer of milk The fat globules are small, so that the milk "creams" slowly, which is an advantage for cheese-making. For beef the Ayrshire is of little account. There is a general consensus of opinion that the male calves are neither worth rearing for beef nor feeding for veal, although the quality of the finished steer is fairly satisfactory.

Ayrshires are largely kept in Canada and the United States, Sweden, South Africa, and New Zealand.

The Jersey (Plate 7, B) which is believed to be of French origin, has existed on its native island for centuries, and no foreign blood has been permitted to be introduced since 1763. A mild and equable climate and a fertile soil render the island very suitable for intensive dairying.

The Jersey is a small breed, the average weight of mature cows being probably under 7 cwt. The colours are fawn, silver grey, brown and brownish black, either whole or more or less broken with white, and with a characteristic "mealy" ring round the muzzle. The horns are small, with an inward curve, and the head is lean and refined, with the orbits very prominent. The cows are generally very "wedge-shaped" and thin-fleshed, and have an angular but refined and deer-like appearance; the skin is thin and the hair short; the udder is generally both capacious and symmetrical, and the milk veins very prominent. The quantity of milk produced is fair, perhaps good, in relation to the size of the animal. The butter-fat content is very high, averaging somewhat over 5 per cent. The fat globules are large and the milk is deep in colour, so that it creams rapidly and has a rich appearance. As a beef animal the Jersey is of no account.

Under favourable conditions of climate and food the Jersey is greatly prized as a butter cow, and she is also a commercial success where there is a sufficient demand, at a special price, for a high quality of milk. The Island herd book was established in 1866 and a separate English register in 1879. Abroad the breed is most numerously represented in the United States, Australia, and New Zealand.

The Guernsey (Plate 8, A) under which name are included the cattle of Guernsey, Alderney, Sark, and Herm, is a larger, less refined, and (in keeping with the less favourable climatic conditions of its home) a more robust breed than the Jersey. The colours are various shades of fawn-red, yellow, and brown-the first predominating; white markings are more frequent than in the Jersey. A light or buff-coloured muzzle is preferred, but dark noses are fairly frequent. The skin is of a very deep yellow colour, and the deeper shades are preferred.

Compared with that of the Jersey, the general conformation is rougher. The shoulder is more loosely attached, the wither is thicker, and the thighs rather fleshier. A certain amount of slackness of back and a prominence at the tail head are common faults. The Guernsey is a fairly deep milker, good herds averaging perhaps 700 gals. The butter-fat content of the milk is slightly lower than that of the Jersey-probably about 4.8 per cent, on the average-but the milk and butter are richer in colour than those produced by any other breed. The Guernsey grows rather longer and reaches maturity more slowly than the Jersey, which latter is often bred from at a very early age. The bulls, as is common in the case of the other small dairy breeds, are treacherous and troublesome to handle, though here, again, the Jersey is the more extreme type. For veal production Guernsey calves have some little value. The steers, however, mature very slowly and are of poor quality, with an objectionably deep yellow-coloured fat. Few steers are reared. The Guernsey is fairly hardy, and thrives in quite exposed situations without exceptional precautions as to shelter or food. There are both Island and English herd books. The breed has been largely exported, the United States being the principal market.

The Kerry (Plate 8, B) and the **Dexter** are the only remaining native breeds of Ireland. The Kerry is an ancient breed, small in size, neatly built, active and hardy, standing on rather long legs and of distinctly dairy type. The colour is black, with frequently a little white on the udder of the cow. The horns are of medium length, rather thin, and curve upwards. On poor hill pastures the Kerry is a commercially useful breed, and gives a creditable yield of milk of rather better than average quality.

The Dexter is of whole black or whole red colour, and of a distinctly dwarfish conformation, with short legs and a heavy head. The general build is compact and blocky, and the breed is a dual-purpose one. The Smithfield figures show that the breed attains rather less than 60 per cent, of the average Shorthorn weight. Crosses with Angus and other beef breeds make excellent small butchers' cattle. Monstrous calves, of the so-called "bulldog" type, are much commoner than in other breeds. Indeed, the Dexter appears to be a heterozygous form, producing, when bred pure, 25 per cent, of long-legged "Kerry type" calves, 50 per cent, of true Dexters, and 25 per cent, of "bulldog" calves.

The Old Gloucestershire is a breed of dairy cattle, brownish black in colour, with a white line extending along the middle of the loin and rump and over the tail head. It has been preserved in very small numbers, but a society was formed in 1919 with the object of reviving it. So far no very great progress can be recorded.

COMMERCIAL TYPES AND CROSSES

There is agreement among livestock authorities that the quality of British cattle, and indeed of British livestock in general, suffers harm from the large amount of indiscriminate crossing that takes place. Some improvement in the situation has occurred since the initiation of bull licensing schemes, under which all young bulls are required to be presented for inspection to officials of the Agricultural Departments. Those that fail to exhibit a reasonable standard of merit are refused licences and have to be either castrated or fattened for slaughter within a specified time. The system was revised in 1945 with the aim of encouraging systematic breeding. A dairy licence is granted only to a bull whose dam and paternal grand-dam have yielded stipulated quantities of milk under official milk-recording. A beef licence is granted only to animals belonging to a recognized beef breed and of good individual conformation. General licences will continue to be granted only until a sufficient supply of dairy and beef bulls becomes available.

The following are some of the better types of cattle that are available in quantity, with the main breeding areas in each case:

1. *Store Cattle suitable for Fatting at Young Ages :* Grade or purebred Herefords from the Welsh Border area and from Ireland; grade Aberdeen-Angus and crosses between Angus and Beef Shorthorns from the semi-upland areas of Scotland and from Ireland; Devons from Somerset and Devon.

2. *Stores for Fattening at Intermediate* Ages : Crosses between Hereford or Aberdeen-Angus bulls and cows of Dairy Shorthorn type from many areas; crosses between Hereford bulls and Welsh Black cows from semi-upland areas in Wales; South Devons and Sussex from their respective home districts; the fleshier types of Dairy Shorthorns from Ireland and northern England; Friesian steers from lowland areas generally; crosses between White Shorthorn bulls and Galloway cows (Blue Greys) from south-west Scotland and northern England.

3. *Stores for Fattening, mainly on Grass, at Older Ages :* West Highland and Shorthorn-Highland crosses from the Scottish Highlands; Galloways from south-west Scotland, Cumberland, and Northumberland; Welsh Black from the mountain regions of Wales.

4. *Dairy Heifers :* Ayrshires from Lanarkshire, Ayrshire, Dumfries, and Wigtown; Dairy Shorthorns from Cumberland, Westmorland, Yorkshire Dales, etc., from Somerset and Wiltshire, and from Ireland; Jerseys, Guernseys, and Friesians from lowland areas generally.

5. *Beef-type Calves :* Single-suckled beef calves are produced mainly in the hill and "marginal" districts of Scotland, northern England, and Wales, and are commonly sold at special auctions in October. The common types are Welsh Black, Herefords, and crosses between the two; Galloways and "Blue Greys" (from Galloway cows by a White Shorthorn bull); West Highland and Shorthorn crosses; and Angus and Angus-Shorthorn crosses. Another useful type includes "colour-marked" calves, mainly crosses between Dairy Shorthorn and Friesian cows by Hereford or Aberdeen-Angus bulls. Such are usually sold at the age of a few days.

BREEDS AND NUMBERS OF LICENSED BULLS

The relative numerical importance of our breeds is shown in the following table, which gives the numbers of bulls licensed in (1) the year about the beginning of the war and (2) the last year for which figures are available :

Breed	England and Wales		Scotland	
	Year ended 31.3.39	Year ended 31.3.53	1939	1952
Aberdeen- Angus	783	478	1,395	1,345
Ayrshire	554	2,365	3,830	3,591
Blue Albion	32	11		
British Friesian	3,086	7, 110	219	490
Devon	1,128	656		2
Dexter	16	23		2
Dun and Belted Galloway	6	6	232	18
Galloway	241	76		264
Gloucestershire	3	4		
Guernsey	1,544	1,379	10	8
Hereford	2,351	1,524	6	132
Highland	4	6	42	81
Jersey	498	661	18	46
Kerry	14	13		
Lincoln Red Shorthorn	1,300	1,045		1
Longhorn	5			
Park (British White).	14	7		1
Red Poll	536	542	4	17
Shetland	2		20	7
Shorthorn	23,80,7	13,290	1,272	814*
South Devon	440	531		
Sussex	238	157		
Welsh Black	440	412		1
Crossbred	34	1	2	...
Total	37,175	30,297	7,050	6,818

*Including 177 dairy types and 3 poll shorthorn

In Northern Ireland, the numbers licensed in 1952 were: Shorthorn and non-pedigree Shorthorn, 1262; British Friesian, 223; Ayrshire, 131; Aberdeen-Angus, 121; Hereford, 93; and Galloway, 23, with very small numbers of other breeds.

PLATE 5

A. RED POLL COW

B. SOUTH DEVON COW

PLATE 6

A. WELSH BLACK BULL

B. BRITISH FRIESIAN COW

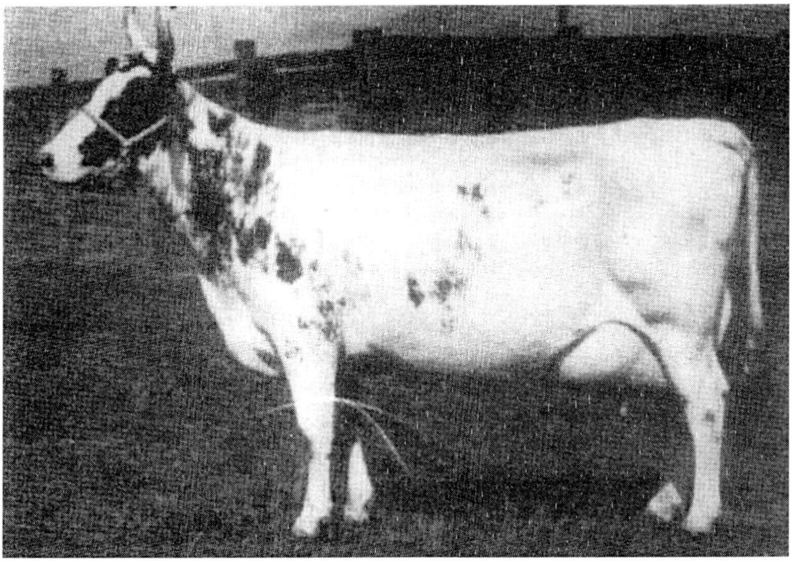

A. AVRSHIRE COW

B. JERSEY COW

PLATE 8

A. GUERNSEY COW

B. KERRY COW

CHAPTER IV

BREEDING AND REARING OF CATTLE-MILK AND BEEF PRODUCTION

THE age at which heifers should first be mated depends on their breed, on their individual characteristics, and on the system of feeding and general management to which they have been subjected during the earlier part of their lives. Females of the early maturing beef breeds, such as the beef Shorthorn, are sometimes, if they be well grown, mated at fifteen months of age, so that they calve down at two years old. Jerseys are frequently put to the bull quite as early. With breeds like the Dairy Shorthorn-which are about average in rate of growth and development, and are suited to average conditions of food and housing-the ordinary age at first service is about twenty-one months, bringing the first calf at thirty months. It is to be noted, however, that where there is a fixed breeding season-as, for example, in pedigree beef herds, cheese dairies, etc.-the choice will fall to be made as between mating at about fifteen months or twenty-seven months so that the heifer may calve in proper season either at two years or at three years old. Again, in whole-milk selling farms it is usually necessary to have a large proportion of the heifers calving in autumn, in order to secure a uniform output of milk throughout the year; and on this account the age for mating may be varied from that which would otherwise be chosen. Animals of the later maturing breeds (Galloway, Welsh, etc.) should not ordinarily calve before they are three years old, and on the poorest Highland grazings, West Highlanders are sometimes not calved down till four. Heifers have been known to take the bull, and to conceive, at as early an age as four or five months, so that they must be kept apart from bull calves from this age until they are considered sufficiently developed to be served.

In view of the many variable factors no precise statement can be made about the optimum age for first mating, but two general points must be borne in mind. First is that gestation makes comparatively small demands upon the heifer's system, whereas lactation, in the case of deep-milking animals, makes very heavy demands. Hence, an animal that is producing a the considerable amount of milk at the age when rapid growth should still be going on, requires a highly concentrated ration if the combined needs for maintenance, growth, and milk production are all to be met in full. If they are not met, the animal will be stunted permanently.

On the other hand there is a widespread belief, which has some support from experimental evidence, that delay in mating may pre-vent the full development of the inherent milking capacity.

Bulls of the earliest maturing breeds may begin to be used at ten or twelve months old; those of average development at about fifteen months. Sixty cows may be put as a maximum allotment for a bull of two years old or over, but much depends on the length of the breeding season, which may extend over two or three months only or may cover most of the year. Breeding bulls must be kept in active and thrifty condition by arranging for sufficient regular exercise, and by so adjusting the ration that the animal is kept in healthy condition without becoming fat.

The ordinary duration of life in cattle is twelve to fifteen years and there are many records of individuals surviving to eighteen or twenty. The average useful life of the dairy cow is very much shorter owing to such troubles as sterility, mastitis, and tuberculosis. The average milking life is about four lactations.

The average period of gestation is 280 days. Cestrus generally begins to occur at from five to eight weeks after calving, so that it is possible for a cow to bear successive calves at intervals shorter than a year. Where all-the-year round breeding is carried on the period between successive calvings, however, may average about thirteen months. When the calves are wanted at a particular season-e.g. March to May in the case of commercial beef-breeding herds-the number of living calves may reach 90 per cent, under good conditions and in the absence of any exceptional factor, such as contagious disease. Under the poorest conditions-e.g. on mountain grazings-2 weaned calves from every 3 cows may be regarded as a satisfactory result.

The interval between successive "heats" varies a little, the av-erage being about twenty days. Females in good condition come in "season" at all times of the year, but most regularly in spring and summer, and least regularly in late autumn and early winter. The length of the heat period is on the average about eighteen hours, but it varies between six and thirty hours. (Estrus is shortest in mid-winter when it may easily pass unobserved. In expected to conceive to the first service. Twin calves are fairly rare-rather above 1 per cent, in dairy herds and about half of 1 per cent, in beef herds. Fertility in the bull depends upon the number and the activity of the sperms, and these vary over a wide range. Bulls may be infertile, either temporarily or permanently.

Calf Rearing : The first period of the calf's life-extending to perhaps six months, when under natural conditions it would be receiving its dam's milk-is one requiring somewhat special consideration. The methods of feeding and general management vary greatly according to economic conditions for example, as the local value of milk and milk products, the cost of labour, the rent of the land, and the probable commercial value of the calf itself.

Certain general principles must, however, be observed under all conditions if ailments and deaths are to be avoided. These are :

1. The calf's powers of resistance to the commoner infections -those that cause white scour, joint evil, and pneumonia-are greatly reduced if it be deprived of the colostrum of the dam. The main explanation is that the colostrum contains "immune bodies" that confer protection against the particular strains of *Bacillus coli* that happen to be prevalent in the premises at the time when the calf is born. In cases where a cow has died in calving or has failed to come into milk immediately, the calf's chances of survival are reduced unless some substitute is provided or exceptional measures are taken to prevent infection. Colostrum from another cow in the same herd (either recently calved or approaching calving) is the best substitute for that of the calf's own dam. The maintenance of a stock of colostrum in deep freeze may also be a useful insurance.

2. Colostrum can be a rich source of vitamin A (and carotene); hence, the cow, for some weeks before calving, should have a reasonable supply of green food, low-temperature silage, etc. The same materials will provide a stock of vitamin E, a deficiency of which is the cause of muscular dystrophy. Another point is that, whereas colostrum forms a soft curd, that produced from the milk of a

"stale" cow gives a hard curd. Such milk, if given during the early days of the calf's life, should be diluted with warm water.

3. In order to prevent infection through the open navel, this should be dressed daily with iodine until it dries up.
4. Calves when housed should be kept singly or in small lots in clean, well-littered pens, so as to restrict the spread of disease.
5. Pail-fed calves should have at least three meals of milk daily for the first ten to fourteen days.
6. A calf on an ordinary allowance of milk will drink upto a gallon of water per day, and water should be freely offered.
7. Easily digestible concentrates and hay of the best available quality should be offered earlier than is the usual practice-from the age of about a fortnight.
8. At the first sign of digestive disturbance milk should be with held for twenty-four hours, and milk diluted with water should be fed for the following day or two. If there are signs that hard curd is being formed in the stomach, an ounce of sodium citrate solution (15 per cent.) should be added to each gallon of milkfed.
9. The general level of feeding during the first few months of the calf's life should be sufficient to support full growth. The consumption of a large amount of coarse fodder, induced by a shortage of milk or concentrates, produces a characteristic lean-backed and pot-bellied appearance.

Veal production, as being a matter rather apart from rearing, may be dealt with first. Surplus calves of the smaller dairy breeds, such as the Ayrshire and Jersey, are frequently slaughtered within a few days of birth. Such animals are, however, of very little value, bringing frequently no more than a few shillings a piece. In some cases the whole carcass is used in the manufacture of dog-biscuit, etc., while in others part is used for manufactured products, such as veal loaf. Calves of the Friesian or Dairy Shorthorn type-or say, Ayrshire-Shorthorn crosses-are, on the other hand, quite suitable for the production of good veal. The average age for marketing may be taken as six weeks, but a good product may be secured at any age from four to twelve weeks. Later the flesh takes on more or less of the character of beef, and the price obtained is less. In the meat trade the term "Bobby" is used to describe a calf of less than three weeks of age, and therefore, necessarily of somewhat inferior quality. The term "Runner Calf is applied" to an animal which, whether by reason of excessive age or the method of feeding employed, has lost the pale colour of flesh that is found in a true veal calf. In order to produce a high quality of true veal the animal should, during the short period available, be kept fattening rapidly, and to this end it should be closely confined and heavily fed. In order to prevent its taking exercise it is generally placed in a small and partially darkened pen, which must be kept clean and regularly littered. Commencing with about ½ gal. daily of colostrum (the first milk of its dam) divided into four meals, the calf should, by the end of the first week, be receiving i gal. thrice daily. This is still further increased, until by the end of the period the ration may be anything from 2 to 3 gals, daily, according to the animal's size. With such treatment a calf weighing 80 lb. at birth may be expected to make a steady live-weight increase of 2 lb. per day, and will thus, at the end of six or seven weeks, scale about 170 lb. live weight and yield about 100 lb. of veal. The quantity of milk required to produce 1 lb. of live-weight increase, when the calf is receiving a full ration, rises from about 8 lb. at the beginning to about 12 lb. at the end, and averages perhaps 1 gal. (10 lb.).

When butter-fat has a high commercial value, as is usually the case, fair results may be obtained on a ration consisting, to the extent of as much as a half, of separated milk plus a cream substitute (see *infra)*. Veal calves are not castrated.

The methods of rear that are adopteding for animals destined for beef, for the dairy or for breeding, may be classified somewhat as follows: (1) The *natural* system, in which each cow suckles her own calf; (2) the *semi-natural,* where a cow suckles two or more calves during a lactation period; and (3) the *artificial,* where the calf is pail-fed, usually, in part at least, on some substitute for whole milk.

1. The natural method of rearing ordinarily produces the best calves, and the chief ground for departing from it is its high cost. With deep-milking cattle the method is, of course, not applicable, even when the value of the calf might justify it, because the quan-tity of milk produced is greater than the calf could take. The natural system is universally adopted in the better herds of pedigree beef cattle, where the value of the calf may ordinarily be expected to cover the whole cost of maintaining the cow for the year. In such herds, in fact, the bull calves, if they are intended for early sale, are sometimes

transferred, at the age of five or six months, from their own dams to freshly calved and deep-milking nurse cows, or may be allowed to suck both. In the production of com-mercial beef cattle the system is mainly confined to districts where cheap, rough grazing is available, and where the cows can be kept out of doors and look after themselves for the greater part of the year. Galloway and West Highland cows are suitable for this purpose in bleak districts, and Hereford or Sussex where the climate is less rigorous. The first two breeds will generally be crossed, if conditions permit, with the beef Shorthorn, in order to obtain larger and earlier maturing calves, Angus cows, kept on better land and housed in winter, are sometimes used in a like manner, and the value of the calves (for intensive feeding into "baby beef" is very high. The increased cost of maintenance of the cow is, however, a disadvantage. Commercial beef calves are normally born in late winter or spring, and are usually weaned in October.

2. It may be reckoned that, with careful treatment and suitable additional feeding, good commercial calves suitable for early fattening can be reared on about 150 gals, of whole milk sufficient, that is, to allow a gallon and a half daily for the first hundred days of the animal's life. It is obvious, therefore, that a reasonably deep-milking cow may be made to rear four calves in a season, and a good milker as many as five or six. Even if we allow a longer suckling period, in order to avoid some of the difficulties, and even if we aim at the production of really well-conditioned calves suitable for "baby beef," an average cow of dual-purpose type should be capable of rearing two or three annually. Again, it is possible, by the use of suitable calf meals containing a proportion of dried separated milk, to adjust the number of calves so that each will get 1 to *1¼* gals, per day. A deep-milking cow (yielding say 5 gals, per day after freshening) may nurse 4 calves for the first three months, 4 more during a second period and 2 more in a third, making 10 in all. The essentials for the success of the system are skilled supervision, a supply of suitable calves, and the collection of a herd of cows that not only milk well but accommodate themselves to the scheme of management. Some cows, even when they are quiet to milk, will refuse to suckle other calves than their own, while others show almost complete indifference. Refractory animals should be disposed of or transferred to the dairy.

When not more than two calves are to be suckled, close supervision is necessary only during the early stages. After the first few weeks the cow and calves may be put to grass with the reasonable expectation that all will go well. Milk will at first largely satisfy the calves' appetites, and their intake of grass will tend to keep pace with the development of the paunch. If, however, the cow is to have three calves during the first part of the lactation, if these are to be weaned and replaced by others in the middle of her milking period, and the process perhaps repeated a second time, the supervision must be close and constant. Here it is necessary to keep the calves in pens or paddocks and to bring in the cows twice a day to suckle under the eye of the herdsman.

The purchase of young calves on the open market involves the danger of introducing disease-particularly white scour-and therefore, such should be kept apart from the rest of the herd for a week or two. In view of what has been said above on the importance of colostrum; it is generally unwise to buy calves that are less than ten days old, unless the -buyer has a freshly calved cow available.

In some cases white scour, which is generally due to invasion of the system by some *form-of Bacillus coli,* may cause heavy loss. The point has already been made (see p. 105) that a cow passes on to her calf, in the colostrum, resistance to the types of *Bacillus coli* that happen to prevail in the place where she is kept. But a calf that is moved to another farm will probably be exposed to different types of the organism, so that white scour may occur despite ordinary care in feeding and sanitation. Dosage with sulfaguanidine is very effective both as a preventive and as a cure. As a preventive, a dose of 2 gm. may be given shortly after the first feed of colostrum, with a second some six to eight hours later. Similar doses should be given morning and evening for the following two days. Affected animals may be treated by dosage at four- to six-hour intervals. The first dose should be 6 to 7 gm., the second 4 to 5 gm., and the subsequent ones 3 to 4 gm., varying with the size of the calf. Dosage should be continued until the scouring ceases.

When the cows are made to suckle successive batches of calves, weaning is necessarily both sudden and abnormally early, and hence, every endeavour must be made to induce the calves to eat a good ration of concentrated food as the time for weaning approaches. They should commence with a small amount of nutritious and palatable concentrate before they are three weeks old. A mixture of three parts oats or flaked maize, two parts finely broken linseed cake, and one part of fish meal is suitable, and the quantity should be raised, if possible, to li lb. before weaning, and to perhaps 21 lb. immediately after. The calves should have some good hay from the age of two weeks, and in winter, from the age of three or four weeks, some finely sliced roots. Later on the supply of concentrates may be augmented, upto a total of 3 or 4 lb. per head per day. Calves that go to grass before they are six months old should continue to receive a little concentrate even if the grazing is very good. If the pasture is of only moderate quality the feeding of concentrate should continue till the age of six or eight months.

3. Pail-feeding has, generally speaking, the object of permitting the partial substitution of other materials for whole milk. In pedigree dairy herds, bull calves and sometimes heifers also are fed by the pail on whole milk only, the objects being to obtain the milk records of their dams and at the same time to regulate the ration; but such a mode of feeding is not general. Separated milk, which has half the energy value and contains all the protein and ash of the natural milk, may be available to the farmer at a fourth of the price of the latter or even less. If separated milk is not available on the farm, separated milk powder may be mixed with warm water at the rate of *I* lb. to the gallon. Whey, which has nearly one-third of the feeding value of milk, can generally be utilized only as animal food, and can be used, with a fair measure of success, in the rearing of calves. Commercial " calf-starters" containing substantial amounts of dried separated milk with vitamin concentrates, etc., enable a consid-erable economy in whole milk. Pail-feeding is the normal system in dairy herds and, in fact, generally where intensive farming is carried on.

Whatever substitutes are to be fed ultimately, the calf should have its mother's milk for ten days if possible, and at any rate for some time after it has lost the obvious qualities of colostrum. Whole-milk feeding I should be continued, if at all possible, for two or three weeks more, the quantity fed being increased from 1 gal. at first to 1½ gal. later-more or less according to the size of the calf. After that time substitution may

[1] The milk of certain Channel Island cows is too rich for their calves. In this case it should be diluted with about 25 per cent, of water.

begin, but should be gradual at first, and should be spread over a further period of two or three weeks. Where separated milk is used, the additional food-the butter-fat substitute-need supply only the energy and the vitamins (A and D) that butter-fat provides. It must be in a readily assimilable form, but apart from this it is a matter of indifference whether it consists of oil, starch, or sugar. Fats and oils, though formerly much used, are generally unsatisfactory when used in quantity In crude liquid form they cause digestive disturbances, and in finely emulsified form they are too expensive. Cod-liver oil, formerly given as a source of vitamins A ana D, is best avoided, since it has an antivitamin E effect, and the essential vitamins A and D can be supplied in pure form. Crushed linseed, oats, and flaked maize are all good, the two latter being preferable on account of their cheapness. Oats need not be crushed or ground; indeed, young calves seem to prefer whole oats, and masticate these quite thoroughly. The cereal meals should be used at the rate of about 3/4 lb. to each gallon of separated milk. If possible, the milk should be fed quite fresh, being separated immediately after milking and given to the calves while still warm. Skim milk, while richer in fat than the mechanically separated article, has the disadvantage of being somewhat stale. In any case, it should be raised to blood temperature before being fed and should preferably be in sweet condition. With separated milk and a fat-substitute a live-weight increase of 1½ lb. daily may be taken as normal.

Where separated milk is available in large quantities, good results may be obtained by substituting 2 gals, of this for 1 gal. of whole milk, adding a source of A and D to make good the vitamin requirements. The calves seem to suffer no harm by reason of the very high protein intake which this method implies.

Whey differs from separated milk in that it lacks not only the fat but also the casein and part of the ash. Moreover, it has necessarily developed some degree of acidity. In order to stop the lactic fermentation, the portion of the whey that is not to be fed immediately it is drawn from the vats should be scalded and cooled. It must be reheated to blood temperature before use. Freshly drawn whey is usually warm enough for immediate feeding. As a supplement in this case it is necessary to use something that is moderately rich in albuminoids. A protein equivalent of 20, and a starch equivalent of the order of 70, may be taken as a general guide, and unless fish meal is included, a mineral addition will generally be beneficial. Examples of suitable mixtures are: (1) two parts linseed cake, two parts flaked maize, one part oats, one part fish meal; (2) two parts linseed cake, one part cracked beans, one part middlings, with 1 per cent, each of ground limestone and sterilized steamed bone flour, and ½ per cent, of common salt. It is, however, impossible to make up a whey ration to an energy value equal to that contained in a full allowance of whole milk; the digestive system of the calf is not capable of dealing with so large a quantity of nutrients in the more bulky and less digestible form. An average calf may receive 1½ to 2 gals, of whey with 1½ lb. of the substitute. The rate of growth will generally be about half that obtained with a full ration of whole milk, or about 1 to ¼ lb. per day, but the daily food cost may be reduced to about one-fourth of that of whole milk feeding.

On farms where milk is produced for liquid consumption, separated milk and whey are only exceptionally available. Ordinarily, in such cases, only heifer calves from good cows are reared, and economy in the use of milk must be closely studied. With care and good management satisfactory results can be obtained with some 30 or 40 gals, of milk per head, spread over the first seven or eight weeks of the calf's life. This allowance can be reduced to perhaps 25 gals, by the use of a carefully compounded milk substitute (calf starter). The feeding of a milk substitute is commenced at the age of two or three weeks and the allowance of milk begins to be reduced at about the fifth week. The substitute-usually a mixture of ordinary feeding stuffs of low fibre-content-may either be fed dry or made up with hot water into a gruel. The former is less troublesome where the animal can be induced to eat a sufficient amount, but some calves refuse to do so, when it is usually best to resort to gruel feeding. The components of the mixture should be easy of digestion, the starch equivalent being of the order of 75 to 80, and the protein equivalent about 18. A fairly suitable mixture for dry feeding might be three parts flaked maize, two parts finely broken linseed cake, and one part fish meal, though better results are obtained if some separated milk powder is included. For gruel feeding special "milk equivalents" or "calf starters" composed of dried whey and separated milk, linseed meal, finely ground cereals, etc.-are obtainable. A suitable formula, which is approved for "National Calf Starter," is thirty parts dried whey powder, fifteen parts dried skim-milk, thirty parts linseed cake meal, ten parts feeding wheat flour, ten parts oats, two and a half parts dried grass meal, one and three-quarter parts carbonate of lime, and three-quarters part common salt. Hay of good quality should be offered from the age of two weeks. A salt-lick should be provided from the time that the calf begins to take solid food.

The allowance of milk products, or of'" milk equivalents" (such as the "calf starter" mentioned above), may be reduced after about three months of age and discontinued at four months old, provided always that first quality hay and ordinary concentrates of high digestibility are available. If the hay is of really choice quality, a concentrate mixture with a starch equivalent of about 65 and a protein equivalent of about 12 will suffice. Suitable mixtures are three parts of oats and one part of linseed cake, or five parts oats and one part decorticated earth-nut cake. Hay should be offered *ad llb.* if the concentrate is readily eaten, otherwise the hay allowance should be such that the racks will be empty for an hour or two before the time for the concentrated feeding.

Consumption of hay may be about 3 to 4 lb. at three months of age and will increase to about 6 lb. at six months. The allowance of concentrate should be of the order of 3½ lb., and need not be increased, since the increasing total requirements are met by the increasing consumption of hay.

Where hay of only average quality is available, the concentrate should be correspondingly richer. Perhaps the best recourse in this case is to continue the feeding of material of the "calf starter" type at the rate of about 1 lb. a day, in substitution for an equal amount of oats or other starchy concentrate. Some farmers prefer to feed the starter as gruel, and to give the other concentrate in dry form.

Finally, it should be said that no attempt should be made to adhere to rigid rules and quantities, since there is considerable variation in the adaptability of individual calves to artificial systems of feeding. Close observation must be maintained and adjustments in diet made according to the progress of the animals. The chief signs of inadequate nutrition are a pot belly, a thin back, and costiveness or laxness of the bowels.

From the age of six months the calf is no longer dependent upon milk or milk substitutes, but still requires a moderately concentrated ration, excluding such materials as straw, if it is to make full growth.

The following is an example of a winter ration for a Shorthorn calf of six months old, weighing 3 cwt., and intended for beef at two years old. Such an animal kept in good growing condition should make a live-weight increase of rather more than 1 'A lb. per day. Requirements may be calculated as follows:

1. Dry matter consumption at 2 ½ per cent. of live weight	8.4
2. Energy requirement	
a) Maintenance	2.70
b) Production, 1.3 lb. live-weight increase at 1½ lb. starch equivalent per lb.	2.28
Total energy requirement	4.98
3. Protein requirement (about one-sixth of the total starch equivalent)	lb. P.E. 0.83

	Composition per cent			In Ration		
	Dry Matter	Starch Equivalent	Protein Equivalent	Dry Matter	Starch Equivalent	Protein Equivalent
20 lb. swedes	11.5	7.0	0.7	2.30	1.40	0.14
3 " meadow hay	85.7	31.0	4.6	2.57	0.93	0.14
1½ " oats	86.7	60.0	7.6	1.30	0.90	0.11
1" flaked maize	87.0	84.0	9.2	0.87	0.84	0.09
1 " palm-kernel cake	89.0	75.0	17.0	0.89	0.75	0.17
½ " white fish meal	87.0	59.0	53.0	0.43	0.28	0.26
		Total		8.36	5.10	0.91

The age at which hand-reared calves may safely be turned out to pasture varies with the conditions. If the pasture be of good quality and "clean" *(i.e.* free from husk and stomach worms), calves born in February may go to grass in May. Where husk is very prevalent, cattle are generally kept indoors until they are at least nine months old.

In New Zealand very satisfactory results are obtained, in the case of dairy calves, by grazing, without concentrates, from the age of eight to ten weeks. This is done under a modified system of strip grazing, with the calves moving ahead of the other stock.

In summer, calves after the age of six months will do well on ordinary grass with the addition of perhaps 2 lb. of any of the less fibrous concentrates. Cotton cake should not be used till the animals are more than three months old; after this age it may be used in moderation provided that the cattle are amply supplied with calcium and vitamins as contained in pasture or in a combination of good hay with some green food.

The general scheme of feeding varies according to the purpose for which the particular animal is being reared. If intended to be fattened at an early age it must have a liberal ration with preferably a fair quantity of whole milk in the early stages. Weaning should be late, and should be carried out by easy stages in order that the "calf flesh" may not be lost. At the other extreme, a heifer intended for the dairy is. purposely kept rather lean, the belief being that otherwise she may develop an undesirable tendency to fatten, and may probably fail to reach her full milking capacity. It is always inadvisable, however, to restrict the ration so much as to check the animal's growth.

As regards housing and general management, young calves should be housed, either singly or in small lots, in well-ventilated but not draughty pens, which should be floored with concrete and swept out and freshly littered daily. They will frequently eat their litter if this be not prevented, and the consumption of straw or peat moss may have fatal results. Calves that develop this tendency may be muzzled for the first three or four weeks, a precaution that also prevents their suckling each other about the navel. In summer they should run on fresh and clean grass, and shade during hot weather is a great advantage; but in many districts it is necessary, as a precaution against "husk" or " hoose," to keep the younger animals indoors during their first summer. This applies to such as have not been weaned at the beginning of the grazing season. When heifer calves are to be vaccinated against contagious abortion, the best age is about seven months. Young cattle at grass should not have ac-cess to the contaminated water of open ponds, since such water may be infected with the organisms of Johne's disease. Males should be castrated at about four weeks old.

When the period immediately following weaning has been suc-cessfully passed, the feeding and management become simple. In summer, young cattle do well on ordinary pasture without arti-ficial feeding, the only exception being in the. case of animals that are intended to be marketed fat at an early age-say, before they are two years old. The latter should have an allowance of 2 to 4 lb. of concentrate, unless the pasture is more than ordinarily good. Yearling cattle will require from half an acre to an acre or more of pasture each, according to its quality. In winter, growing cattle may be tied up in stalls, or run in covered courts or partially open yards. In districts where the winter is mild *e.g.* the south-west of England and south-west of Ireland-those of eighteen months or more do best out of doors, if reasonably dry and well-sheltered pastures are available. Store animals will thrive when closely confined, but young ani-mals that are being intensively fed must have exercise, if they are to be prevented from "going off their legs." An open court, or failing this, a covered court with a daily run on pasture, is preferable even in the case of stores, as the animals grow better coats and are much less liable to suffer a set-back, when they come to be turned out in spring. The following are examples of winter rations :

1. For a twelve-months-old store bullock weighing 5 cwt., in an arable district:

	LB
Swedes	40
Oat straw	8
Oats and beans, or compound dairy cake	3

This supplies about 14 lb. of dry matter (equal to 2½ per cent of the live weight), 6.4 lb., starch equivalent, and 0.85 lb. protein equivalent. The maintenance energy requirement is 3.7 lb. starch equivalent, and the surplus, viz. 2.7 lb. starch equivalent, is sufficient for something over 1 lb. live-weight increase per day.

2. For a two-years-old store bullock of 8 cwt., intended for early summer fattening, in a grassland district with a daily run out on pasture :

	LB
Meadow hay	20
Oats, barley, or other cereal	4

This supplies 21 lb. dry matter, 9 lb. starch equivalent, and 1.2 lb. protein equivalent. The maintenance requirement is 5.5 lb. starch equivalent, so that the balance available for production would be 3.5 lb., sufficient for fully It lb. live-weight increase. In actual practice the full ration of hay would be given only for a short period, say from January till

March, the quantity being regulated during the remainder of the winter according to the state of the grazing.

The first essential to any profitable system of milk production is a herd of deep-milking cows. It is true that the actual yields, as well as the profits, depend very largely on the skill of the dairyman in feeding and general management; but no management, howsoever perfect, can prove successful if the cattle have a low limit of production. Hence, everything possible must be done to secure good foundation stock, and, by careful selection and mating, to maintain or improve the average yield. To this end a system of milk recording is essential. Individual cows, under what may be termed ordinary farm conditions, vary in yield from 200 or 300 to 1000 gals.[1] or more, and accordingly may leave a large profit or a very considerable loss. The average yield of the world's dairy cows is probably not over 400 gals., whereas good herds, with a standard of management that is quite attainable under strictly commercial conditions, may reach 800 or 1000 according to breed, and occasionally more. The composition of the milk varies, too, with the breed and with the individual; in particular the percentage of butter-fat, which is the most valuable constituent, may be anything from 3 to 6, without counting very abnormal individuals. There is neither adequate evidence as to the hereditary factors that determine production, nor of the mode of their transmission; but we know, in a general way, that milking qualities are inherited, and hence, milk records, apart from their value as a measure of the economy of production of individual cows, are of the utmost importance in breeding.

An actual record must always be considered in relation to the conditions under which it is made. Obviously, the size of the cow must be taken into account. The economy of milk production, as has previously been pointed out, depends essentially on the ratio between the production ration and the maintenance ration; the latter varies, not in direct proportion to the live weight but with the three fourths power of the live weight. Thus, for example, a 12 cwt. cow will require about 35 per cent, more for maintenance than an 8 cwt. animal, and hence it may be calculated that 800 gals, from the latter would be as profitable a yield as 1080 gals, from the former. Clearly, too, a record must be considered in relation to the general system of feeding. An insufficient ration will limit the production and pre-vent the animal from giving any true account of herself.

[1] A gallon of milk is taken for practical purposes as 10 lb.

Again, age has a well-known influence on yield. From the fourth lactation onwards the cow may be considered as mature, and little increase in yield occurs after this time. Naturally, the yield declines ultimately, but the age at which the decline sets in may vary from eight years to perhaps twelve. A heifer calving for the first time at thirty months old may be expected, in her first lactation, to produce about 75 per cent, of her mature yield, and with her second and third calves about 85 and 90 to 95 per cent, respectively. The butter-fat content of the milk varies but very little with age.

The normal lactation may be taken as lasting about ten and a half months. The cow should have at least six weeks-and preferably eight weeks-rest before calving, and should be in calf again three months after freshening. An abnormally long period of rest previous to calving has the effect of increasing the yield during the subsequent lactation, up to a maximum of about 15 per cent, above the normal yield. Conversely, if a cow is not rested at all, her subsequent yield may suffer to the extent of 15 or 20 per cent.

Delay in mating has the effect of permitting a longer continued flow, and a cow that does not conceive at all will ordinarily give about 30 per cent, more milk than one which conceives at the normal period of three months after the previous calving. Under the National Milk Records Scheme (England and Wales), the standard lactation period is 305 days-i.e., the lactation yield is taken as the amount produced in 305 days from calving. A long period of rest, particularly if the cow be fattened, may have the further result of increasing the butter-fat content of the milk during the first few weeks of the sub-sequent lactation.

Finally, the season of calving has an important influence on the yield. The change from winter to summer feeding usually stimulates the flow, whereas the generally poor quality of autumn pasturage has the opposite effect. This effect is greatest if the change occurs at the time when the yield is beginning to decline, say three or four months after calving. It has been found that a cow calving in December may yield 10 to 15 per cent, more than another of equal real capacity calving in June; but, obviously, no generalisation can be true for all conditions, and the more skilful the winter feeding the less will be the effect of season of calving. Thrice-daily milking produces a higher yield than the usual twice milking, and this should also be remembered in interpreting records.

Even when all these factors have been allowed for, an actual record does not necessarily reflect accurately the inherent milking capacity of the cow. The level of nutrition in the herd and the degree of skill brought to bear in its management have very considerable effects on yield.

Many attempts have been made to estimate milking capacity on the basis of short tests-of, say, seven days' duration. It has been shown that with cattle that are fed and managed in the ordinary way there is actually a very high correlation (about 0.85) between the maximum daily yield, at, say, six or eight weeks after calving, and the total yield in a normal lactation. The relationship is, on the average, such that if the maximum daily yield of a winter-calving cow is multiplied by 200, there is obtained a very fair approximation to the normal lactation yield. For summer-calving cows the factor would be about 185. Thus, an average cow, producing, say, 3 gals, a day at her best, might be expected, if she calved in winter, to reach 600 gals, in a normal lactation. Since, however, the correlation is not absolute, the rule is not strictly applicable to the poorer milkers or those that are much above the average. The difference from the av-erage should be multiplied by 0.85, the coefficient of correlation, and the product again multiplied by the lactation factor. Thus, if we assume that 600 gals, is an average yield, a 5 gal. cow, calving in winter, should produce in a lactation (3 x 200) + (2 X 0.85 x 200) = 940 gals. Similarly, a winter-calving cow, yielding 2 gals, at the top of her flow, might be expected to produce (3 x 200) -(1 x 0.85 X 200), or 430 gals., in a normal lactation.

Short-period tests, in spite of their inherent value, are of very little 'use in practice. The difficulty is that it is possible, by spe-cial feeding, to stimulate the flow for a short period, as well as temporarily to raise the butter-fat content. Even yearly records are open to objection in cases where the cattle are carefully rested and prepared, or where mating is delayed beyond the normal time. The average production over a series of three or four years is by much the best measure of value. Indeed some of the breed Societies now stress the importance of this by setting up the "10,000 gal. cow" as the ideal-i.e. a cow which yields an average of 1000 gals, per year over a productive life of ten years.

Milk records, if they are to receive general recognition, must obviously be vouched by an independent authority. In some systems the milk is weighed and tested at regular intervals by the official recorder, but this method is costly if the visits are frequent enough to give an accurate estimate of the yield. The better system is to make the owner responsible for weighing and recording the yield at each milking, and to have the milking inspected, the weights checked, and the milk tested for butter-fat content by inspectors making surprise visits.

FEEDING FOR MILK

The feeding of dairy cattle, if maximum production is to be made possible and waste prevented, is necessarily a somewhat complicated business. Sufficient energy, protein, minerals, and vitamins must be supplied to provide for maintenance and something approaching the quantity of milk that the cow is capable of yielding. The maintenance requirement is 0.6 lb. of protein equivalent and 6 lb. starch equivalent per 1000 lb. live weight. A gallon (10 lb.) of average milk contains 0.35 lb. of protein, so that the protein requirement, per gallon of milk produced, cannot be under this figure. In fact, good results have been obtained with an allowance of 0.45 lb. per gal., but a somewhat higher rate is probably advantageous in maintaining production during the later stages of lactation. The accepted standard is now 0.55 lb. protein equivalent per gallon of average milk (3.75 per cent, butter-fat).

The total or *gross* energy value of a gallon of average milk (3.7 per cent, butter-fat) is about 3200 Cals, If, therefore, we were to take Kellner's value for the pound of starch equivalent viz., 1071 Cals of net energy it would be necessary to feed 3 lb. $^{3200}/_{1071}$ of starch equivalent for each gallon of milk produced, for the cow must have enough net energy in her food to produce the gross energy in her milk. The value of 1071 Cals. of net energy per pound of starch applies, however, to fattening oxen, and it has been shown that the milking cow is distinctly more efficient as an energy transformer than the fattening steer. The probable explanation of this fact is that the cow has a direct outlet in her milk for protein and carbohydrate, and is not under the necessity, as is the fattening animal, of trans-forming its whole surplus energy into the form of fat. The difference in efficiency is considerable, amounting to fully 25 per cent. Expressed differently, the pound of starch equivalent yields 1360 Cals. in the form of milk as against 1071 Cals. in the form of body fat.

The quantity of starch equivalent required per gallon of milk of average composition is thus, $^{3200}/_{1360}$ or 2.35 lb. approximately. The energy requirement per gallon will, of course, depend on the composition of the milk, notably on its fat-content. Moreover, cows that are fed in excess of their calculated requirements maintain relatively larger yields, in the later stages of lactation, than those which receive bare theoretical allowances. The optimum level of feeding thus depends on economic conditions, particularly on the relative prices of feeding stuffs and milk. The table below embodies the recommendations which are generally accepted, though some individual authorities regard the figures as somewhat too high for average conditions.

Fat-content of Milk (per cent.)	Cals. per Gallon	Starch Equivalent lb.	Protein Equivalent lb.
3.5	3170	2.40	0.52
3.75	3250	2.50	0.55
4.0	3430	2.60	0.58
4.25	3510	2.70	0.61
4.50	3600	2.80	0.63
4.75	3680	2.90	0.66
5.25	3820	3.10	0.72

In the case of the cow giving milk of average composition, and where ordinary cakes and meals are to be used as the production ration, the quantity required will be about 3 ½ lb. per gal. This implies a starch equivalent of 71 per cent, and a protein equivalent of 16 per cent. Such a ration will contain fully 3 lb. of dry mat-ter. As is pointed out in the chapter on nutrition, the cow's food capacity generally lies between 2½ and 3 lb. of dry matter per 100 lb. live weight per day; 3 lb. of dry matter per hundredweight (112 lb.) is a good general figure. Hence, if we are dealing with cows of 11 cwt. live weight, with yields running upto 5 gals., maintenance requirements must be met by a ration containing not more than 18 lb. of dry matter (33 lb. total minus 15 required for production). The following are the examples of rations supplying the necessary minima of energy and protein (7.1 lb. starch equivalent and 0.71 lb. protein equivalent) for maintenance in not more than the bulk indicated :

a) 20 lb. good hay (medium hay is generally given a starch equivalent of 32, very good hay one of 40. A starch equivalent of 35 would meet the case),

b) 40 lb. swedes or mangolds with 14 lb. medium hay.

c) 30 "tare-and-oat" silage with 10 lb. medium hay.

d) 60 swedes, 6 lb. oat straw, 6 lb. medium hay.

The first is suitable for grassland farms, the second and third for mixed farms; the fourth can be successfully used only in districts where the roots and straw are known to be of at least average quality, for there are many places in which these foods fall short, in nutritive value, of the figures given in the tables. On the other hand, in areas where roots and straw are exceptionally nutritious-e.g. in north-east Scotland-the last ration might provide not only for maintenance but also leave some surplus for production.

As regards the production ration, this ordinarily consists of a mixture of three or four distinct materials. The following are suitable standard mixtures to be fed at the rate of 3½ lb. per gal :

a) Two parts maize meal, one part crushed oats, one part bean meal, two parts palm-kernel cake, one part decorticated earth-nut cake.

b) Two parts maize gluten feed, one part rice rneal, one part barley meal, one part linseed cake, one part decorticated cotton cake.

c) One' part coconut cake, one part decorticated earth-nut cake, one part linseed cake, two parts palm-kernel cake, two parts rice meal, one part maize gluten feed.

In the choice of concentrates, apart from questions of price, due regard should be paid to the type of maintenance ration. If this is costive in tendency the production ration should be laxative and *vice versa.* Coconut cake, linseed cake, and decorticated earth-nut cake are laxatives, while cotton cakes, dried grains, and most of the starchy foods have the opposite tendency.

Since 1939 individual "straight" feeding stuffs have been in very irregular supply, and most fanners have made use of compound cakes, and meals.

Dairy compounds are of two broad types-viz. *balanced* and *balancer,* the latter for use, in conjunction with horne-grown cereals, in specified proportions.

It will be clear that a scheme based on maintenance and production rations, such as those indicated above does not meet the case of heavy milkers, viz. of cows of average appetite giving more than 5 gals, per day. While the proportion of cows that actually give 5 gals, or over on winter feeding is small, there is no doubt that many more are inherently capable of doing so. A very partial solution of the difficulty is to increase the concentration of the production ration in such degree that 3 lb. will provide the necessary nutrients for 1 gal. This implies a starch equivalent of 83 and a protein equivalent of 20 per cent. Such a mixture might be composed, for example, of equal parts flaked maize, palm-kernel cake, bean meal, coconut cake, decorticated earth-nut cake, and crushed linseed. But a more satisfactory method is to reduce the" bulk of the maintenance ration by substituting concentrates for hay. It has been found that the coarse fodder can be reduced to 8 or 10 lb. without harm to the health of the cow. The maintenance ration may be made up with the lighter type of concentrate, such as a mixture of two parts bran to one part flaked maize. Thus 10 lb. of good hay, 4 lb. bran, and 2 lb. flaked maize would provide the maintenance requirements of a 1200 lb. cow, and leave room for a 7-gal. production ration. The feeding of still heavier milkers becomes something of an art, involving careful study of the taste, appetite, and reactions of the individual animal.

The object of rationing according to yield is, of course, to ensure, without waste of food, something approaching the maximum yield of milk of which the cow is capable, while enabling her to maintain herself in good thriving condition. Cases of individual cows becoming too fat or too lean do frequently occur in rationed herds, generally because due regard is not paid to variation in the composition of the milk. In such cases the ration should be altered at discretion.

An important point in dairy-cow management is that the cow should be brought into fit bodily condition and accustomed to a fairly substantial ration of concentrates, before she calves. This process may commence (with 3 or 4 lb. of cake per day) six weeks before the calving date, and the ration may be raised to 8 or 10 lb. for the last fortnight. A mild purgative should be given, when it appears that parturition is imminent. Only a light ration should be fed for the first few days after calving; but thereafter the concentrate should be rapidly increased day by day until it exceeds, by 2 or 3 lb., the allowance corresponding to the actual yield. Excess should continue to be fed so long as the yield shows a tendency to rise, which may be for as long as six or eight weeks. Thereafter, the ration should correspond to the actual yield.

The mineral-content of the ration requires special consideration in certain cases. The moderate milker receiving a full maintenance ration will normally get enough mineral matter, at least if the bulky foods have been produced on land that contains sufficient lime and phosphates. The heavy milker, on the other hand, will receive a large proportion of her food in the form of cakes and meals, most of which are deficient in minerals, particularly calcium. The addition to the production ration of 3 per cent, of a mineral mixture will be desirable in such cases. One part powdered chalk, two parts common salt, and one part sterilized steamed bone flour form a mixture which has given satisfactory results. Blocks of mixed minerals, which are obtainable commercially, may, as an alternative, be fixed in the mangers, and the cattle allowed to lick them at will.

The composition of the milk is in the main independent of the feeding, but rations that are low in calcium tend to depress the content of solids other than fat. Chronic under nutrition, produced by the continued use of rations of inadequate energy value, seems to have the same effect. Some materials, such as maize gluten feed, coconut cake, palm-kernel cake, and earth-nut cake have a slight and sometimes only temporary effect in the direction of increasing the butter-fat content. On the other hand, there is some evidence that rice meal has a depressing effect on the fat content. The matter is, however, not fully understood. In any case, a high oil-content in the ration is undesirable, and it should not exceed 5 or 6 per cent. The character of the butter-fat is markedly influenced by feeding-linseed cake, for example, produces a soft oily butter, while bean meal, coconut cake, and palm-kernel cake produce a firm fat. The colour of milk, due largely to its content of carotene, is also greatly influenced by feeding, although, as previously pointed out, it varies with the breed and individuality of the cow. Some breeds, notably Guernseys and Jerseys, pass on carotene in their milk, which accordingly has a deep creamy colour. Cows of other breeds, such as the Ayrshire and Friesian, convert a much larger proportion of the food carotene into the colourless vitamin A. The green leafy portions of plants have a high carotene-content and hence, pasture grass, kale, etc., impart to milk a desirable creamy tint. The carotene of grass, etc., is fully preserved by artificial drying or by low-temperature ensilage, but is largely destroyed in the ordinary process of haymaking. The point is of importance, since milk of high carotene-content has a high vitamin A potency. The amount of food carotene or vitamin A required to maintain the full potency of the milk is not large. It can be provided by about 5 lb. of fresh young grass, by 1 lb. of high quality dried grass, or by 2 oz. of cod-liver oil per day.

Regarding the arrangement of the diet, concentrate may suitably be given just before milking. Roots, beet tops, or silage, which, when given shortly before milking, are most liable to taint the milk, should be given not less than three or four hours before milking[1]; long fodder, which is liable to contaminate the air with dust, should be fed soon after milking. The following is an example of a suitable arrangement of the diet :

5.45 A.M.	One-half concentrate
6 to 7.15 A.M.	Milking
7.15 A.M.	Half root ration
9 A.M.	Half long fodder
10 A.M.	Outdoor exercise in fine weather
2 P.M.	Half root ration
4.45 P.M.	One-half concentrate
5 to 6.15 P.M.	Milking
6.30 P.M.	Half long fodder

The actual times are determined in certain cases by the hours at which the milk has to be delivered.

On the question of the preparation of the food it may be taken as a good general rule that roots should be fed either whole or roughly sliced, coarse fodder in the long state, and the concentrate in dry form. Chaffing may be desirable in the case of animals that waste good fodder. Straw and poor-quality hay are best fed unchaffed, so as to allow the animal to reject the coarsest and least nutritious portions. For production purposes a mixture of cake (in nut form) with meal is better than meal alone. The feeding of wet mash is common in certain districts, but it is a method of very doubtful economy-except, of course, where wet beet pulp or brewers' wet grains are used.

---

[1] Where the cattle spend an hour or two at pasture, such foodstuffs are best fed out of doors; this avoids the risk that the taint will be picked up from the atmosphere of the cow-shed.

A supply of pure drinking water is necessary for cattle in milk, unless the ration contains both an excessive quantity of roots and a considerable amount of water in the form of mash. The normal consumption of a cow yielding 2 gals, of milk per day, fed on a dry ration, and in ordinarily cool weather, may be put at 10 gals, daily. Heavy milkers, in hot weather, have been observed to consume up to 25 gals. The most satisfactory method of supplying water, when the cattle are indoors, is by means of automatic drinking bowls. Failing these, water should be offered thrice daily, the last drink being given as late in the evening as is practicable.

In summer, if cattle are on fairly good fresh pasture, it will generally suffice to give 3 ½ lb. of concentrate for each gallon of milk produced above 2 gals.; or, in case of average pasture during the earlier part of the season, for each gallon above 3 gals. The best grazings, for a few weeks during May and June, can support production upto 5 gals, per day. When pastures get bare, much additional food may require to be given, such as cut green forage, mangolds in early summer or cabbages later, hay, and additional concentrate. Even if grass is abundant its feeding value generally declines from August onwards, and during October it will do little more than provide for maintenance. Deep-milking cows on rough autumn pastures should have their grazing restricted, since other-wise they may consume too much of the bulky and innutritious herbage. An average cow requires from 1 to 1³/4 acres of pasture, according to its quality, when this is managed on the ordinary system, or about ¾ acre under a highly intensive system of management.

Pastures that contain an excessive proportion of legumes *e.g.* white clover or lucerne-are liable to cause bloat. Young cattle, fat-tening cattle, and sheep are liable to the trouble, but it tends to be most severe in the case of cows in full milk. In some cases the feeding of hay or other roughage before the cattle go to pasture is an effective preventive.

Where soiling is practised, the ration of green forage may be from 60 to 80 lb., according to its feeding value and the size of the cow, with the usual allowance of concentrate. High-quality material may be fed in larger amounts, and the concentrate reduced in amount or omitted.

It is worth noting that many summer foods, such as clover, vetches, and ordinary pasture grass, are much richer in protein than most home-grown winter foods, such as roots and hay.

PLATE 9

Milking Stals with Releaser Milking Machine

Hence, the concentrate fed in summer may frequently be so made up as to have a nutritive ratio of the order of about 1 : 6 or 7 and yet give fully satisfactory results. This may permit the use of a mixture of, sharps, maize meal, and palm-kernel cake, which will often be appreciably cheaper than type of mixture.

In all cases feeding should beat strictly regular hours. The cattle should be quietly handled, should be groomed daily during winter, and should have some little outdoor exercise, except when icy roads might lead to accidents, or when the weather is exceptionally inclement The hoofs may requite occasional diessing to pi event their glowing too long.

HOUSING AND HYGIENE

Cow-sheds should have ample air space. Some local authorities prescribe 600 cub. ft. per cow, or 800 if the cattle are continuously indoors. Much important than the actual cubic space are efficient ventilation, good lighting and drainage, and facility for cleaning. The floor should be of non-absorbent material, such as concrete, except perhaps, the front part of the stall, which may be of rammed clay or chalk, in order to prevent injury to the cows' knees. Rubber bricks and composite bituminous materials are now being used with the same object. The size of the stalls must be adapted to that of the cows; for moderately small breeds, such as the Ayrshire, the double stall should be about 6 ft. 6 in. wide and about 5 ft. 3 in. long, exclusive of the through. For average Shorthorns an extra 6 in. each way will be necessary. If yokes are used instead of chains the stalls should be made 6 in. shorter, as cows in yokes stand farther forward. Too much width or length is just as unsatisfactory as too little. In the first case a cow is liable to turn partly round in the stall, and to tread on her neighbour, as well as to soil the litter. The depth of the manure channel should be 10 in. behind the cow and about 4 in. next the passage. A suitable general arrangement Of a cow-house, with dimensions, is shown in Fig. 8.

The atmosphere of the shed should be kept cool and fresh. It has been shown that there is no gain in milk by keeping the cows at a temperature above 45° to 50°F and a high temperature has been shown, in some cases, to depress the butter-fat content of the milk. Even in the coldest weather ordinarily experienced in Britain, free ventilation is unlikely to reduce the temperature to any harmful extent. At the cooler temperature the callle are imicri less liable to chills, and are more weather-hardy when they come to be turned out in spring. An adequate supply of fresh air to each cow may be assured by inserting glared stone-ware pipes in the cow-shed wall, one opposite each standing, at about UK level of" the top of" the manger. The cattle should be kept littered with clean straw, and the floor should be washed as well as swept daily. The lower-part of me walls, which should be smoothly faced with cement or other material, should frequently be brushed down with water.

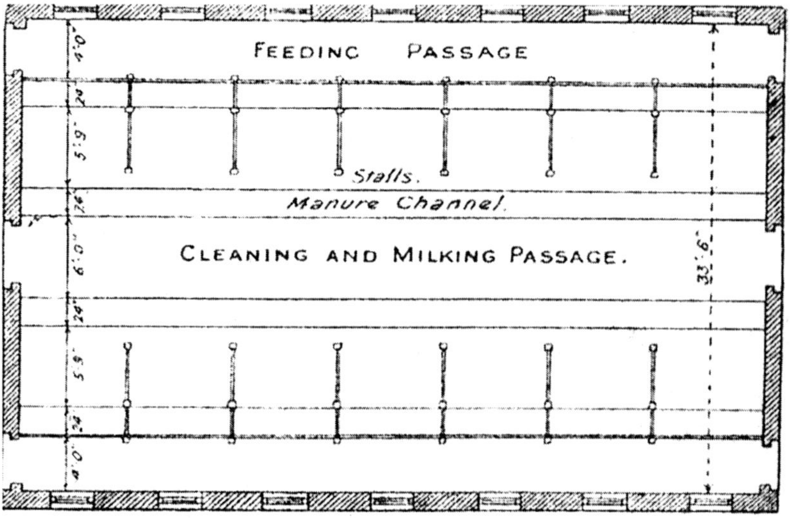

FIG. 8 : Plan of Double shed for 28 Cows. Dimensions are cows of average size. Stalls are 7 ft. wide

The purity of milk, which determines its keeping quality, is judged by its bacterial count and its content of coliform organisms per millilitie, or alternatively by its effect on such reagents as methylene blue or resazurin. These tests depend upon the fact that active bacteria use up oxygen, and that the reagent changes colour on reduction. Milk withdrawn from a healthy udder contains very few bacteria but tends to become contaminated from such sources as unsterilized utensils, the body of the cow, the hands of the milker, dust, etc. As milk is a very suitable medium for bacterial growth, the count tends steadily to increase. The rate at which the organisms multiply is most rapid about blood heat but is less at lower temperatures, and goes on comparatively slowly if the milk is cooled to 50°F Milk from unhealthy cows may be contaminated by pathogenic bacteria within the udder. Milk infected with the various organisms that cause mastitis is, naturally, capable of infecting other cows but is generally innocuous to humans; milk containing the organisms of tuberculosis is highly dangerous to both humans and bovines.

The following precautions should be taken to ensure clean milk production :

1. The hind quarters of the cows should be kept clipped and the animals should be groomed daily, but grooming must not take place immediately before milking as this causes contamination by dust during milking. The udder and teats should be washed before milking, but care must be taken that the cloths used for this purpose do not carry infections, such as mastitis from one cow to another. To sterilize the clothes, hypochlorite solution may be used according to the instructions of the makers. Where this isdone, it is recommended that two pails of solution should be prepared and two clothes added to each. The first pail is washing down, the clothes being used alternately so that one is in the solution for disinfection while the other is being employed on the cow. The clothes in the second pail are used, time about, for giving a final wipe-down. The first "draw" of milk, which contains a good many bacteria from about the teat orifice, should be rejected.

2. The atmosphere of the cow-shed should be kept free from dust. Dust may come in from outdoors or may be produced by handling long fodder or litter before milking. As it is impossible to exclude bacteria from the atmosphere altogether, a hooded pail should be used (where milking is done by hand) in order that a minimum surface of milk may be exposed. Milk should be removed from the cowshed as each cow is milked.

3. Personal cleanliness should be required of the milkers, and washing facilities should be provided. The hands may be dipped in hypochlorite solution before and after milking each cow. Wet milking should be forbidden. People suffering from infectious disease, sore throat, sores, or diarrhoea, must not be allowed to handle milk.

4. All utensils, such as milking pails, etc., should be thoroughly cleansed and sterilized before milking begins. After washing in cold or tepid water, scrubbing in hot detergent solution, and rinsing with hot water, the utensils are placed in the steam-chest and, after sterilizing with live steam for ten minutes after the chest thermometer shows a temperature of 210° F., are left until next required.

5. The milk should be passed through a suitable strainer and immediately cooled, in a clean atmosphere, to as Iowa temperature as the available water supply will permit or, where refrigeration is available, to below 50° F.

6. Careful watch should be maintained over the health of the cattle, and the milk of any unhealthy animal should be excluded from the general supply. It may, after scalding, be fed to pigs.

7. In order to prevent the spread of mastitis, which is a frequent cause of heavy losses, the milk from each cow may be subjected to bacteriological examination. Where this has been done, all the "clean" cows should be milked before the workers pass to the affected cattle. Still better, each section of the herd may have its own group of milkers. The incidence of the disease is highest in the older animals, so, when the condition of the udders is unknown, it is sound practice to milk the old cows last and the heifers first. As mentioned above, it is also helpful to sterilize the udder cloths and dip the hands in water containing hypochlorite.[1] In the case of machine milking, after each cow the vacuum should be turned off and the cluster of teat cups rinsed in clean cold water, drained for a few seconds, and then dipped in hypochlorite solution. After a final shake the vacuum may be turned on and the machine used on the next cow.

Some authorities are chary in recommending the employment of disinfectants because unreliable workers may make up excessively strong solutions and do harm, or continue to use the hypochlorite solution after its activity is exhausted. Where there is reason to anticipate mishandling of disinfectants, it may be better to wash the udders, etc., in clean water and to change the latter very frequently.

---

[1] There is some hope that some more completely effective means of hand disinfection may be found.

It will be obvious that these general rules may be interpreted in various ways. Where the milk is sold as " ordinary" milk, the price obtained will rarely be sufficient to compensate the producer for the installation of elaborate sterilizing and refrigerating plant, or for all the labour that is required to ensure the minimum of contamination. On the other hand, where the highest class of graded milk is aimed at, no detail can be neglected.

The system of open-air dairying, first devised by Mr A. J. Hosier in Wiltshire, is now fairly common on well-drained grasslands in the South. The cows live out of doors all the year round, and are milked in a "bail," which is a movable set of milking stalls with milking machine and cooler built into the structure. The cows receive their concentrate ration while they are being milked and return to the field immediately milking is finished. The system is very economical in labour, two men sufficing for a herd of sixty or seventy cows. The cattle generally maintain excellent health, though the milk yields are probably a little below those of similar animals managed in the usual way.

Another method of management is to keep the cows either in an ordinary cow-shed or loose in yards (when horns should be cut), and at milking time to drive them into a shed with special stalls, where they are first washed and then milked. A suitable arrangement is illustrated in Fig. 9. A parlour with side-by-side milking stalls and a releaser type milking machine is shown in Plate 9.

Such "milking parlours" are being installed in increasing numbers of dairy farms. The cost of fitting out a parlour is much less than that of building a good cow-shed, and existing buildings can often be cheaply adapted as cow-yards. Labour costs are lower and the manure, being made in yards, is much better conserved. The chief difficulty is in rationing the cows, since the period spent in the milking stall is insufficient to enable a heavy milker to consume the appropriate production ration. Many modern yards are, however, equipped with yoked mangers which permit of individual rationing.

MILKING AND FARM DAIRY EQUIPMENT

Four sorts of equipment are needed in the routine operations of milking and milk-handling, whether milking takes place in the cow-shed or in a milking parlour. These are: *(a)* Udder washing equipment and a strip cup; *(b)* a milking machine or hand-milking pails, and cans into which the milk is finally placed to await dispatch from the farm; *(c)* strainers and coolers which are used to minimize bacterial growth in the milk; and *(d)* washing and sterilizing equipment for keeping the utensils in a satisfactory hygienic condition A special room, the daily, where all operations except the actual milking are carried out, is required as an adjunct to the cow-shed or milking parlour.

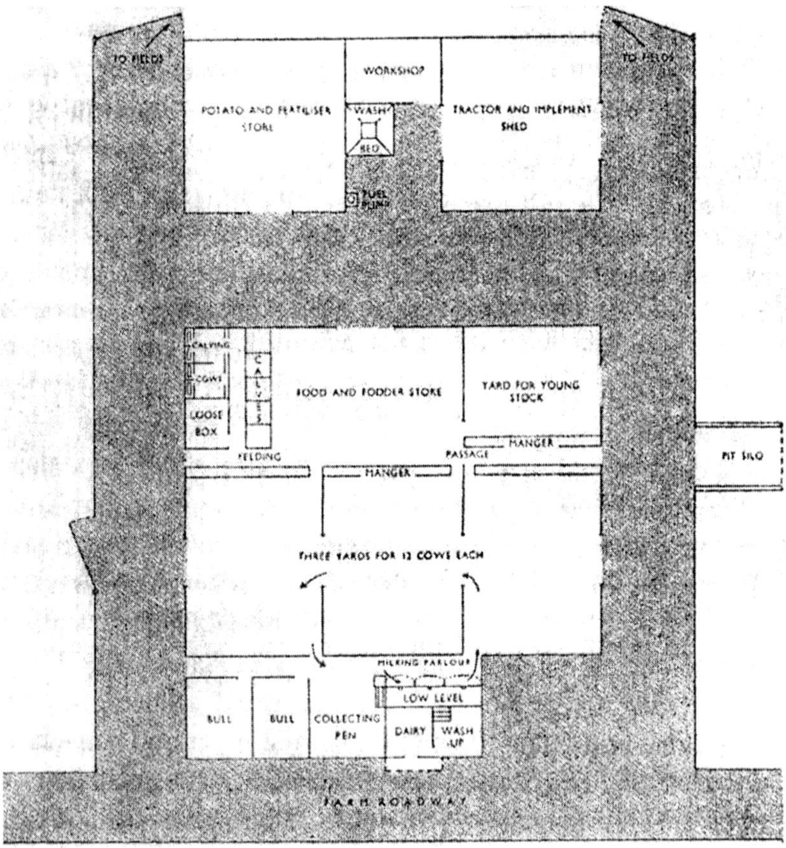

FIG. 9: Arrangement of Yards and Tandem Milking Parlour for Herd of 40 Cows

Milking Machines : All milking machines work on the same general principles and aim to remove milk from the udder as quickly and completely as possible without injury to the animal. Little exact information is available regarding the effects of specific design features on any of these objectives. The milking action of machines results from a constant vacuum (of about 15 in. of mercury) applied to the inside of a rubber sleeve enveloping the cow's teat, possibly aided to some extent by an alternating pressure applied to the outside of the sleeve or liner. The liner is contained in a metal teat-cup and makes a vacuum tight seal to it top and bottom. By connecting the space between the liner and cup alternately to the atmosphere and to the same source of vacuum as operates inside the liner, the liner is made to squeeze and release the teat alternately. During the squeeze phase of the cycle, milk flow from the teat ceases as the liner collapses and the teat is protected from the full force of the continuous vacuum applied to the liner. In the absence of pulsation engorgement of the teat occurs with consequent discomfort to the cow and injury to the tissue. The alternating pressure is controlled, at about 50 cycles per minute, by a pulsator which is often a simple valve mechanism operated by a vacuum motor similar in construction to that of a windscreen-wiper. A vacuum pump, driven either by an electric motor or by an internal-combustion engine, is an essential part of the machine.

In the releaser type of installation the milk from several teat-cup clusters is conducted by rubber milk tubes into a metal pipe, the milk-line, fixed at about head height to the standings. The milk-line conveys the milk to the dairy, where it is brought to atmospheric pressure (from the reduced pressure prevailing in the milk tubes and milk-line) by means of a device known as a releaser. The vacuum-line usually runs parallel to the milk-line and pulsators are fitted at each cow standing. In the more elaborate recorder-releaser plants the milk from individual cows is first intercepted by a counter balanced glass weighing jar and, after milking is finished, the weight of the cow's milk as indicated by the spring balance is noted. Then, by operating a master valve, the milk is drawn from the jar into the milk-line.

While releaser plants are commonly employed in milking parlours, bucket milking plants are normally used for machine milking in cowsheds. Vacuum cocks at each standing are linked by the vacuum-line to a motor-driven pump, and the portable milking units are connected in turn to the cocks. Each unit consists of a teat-cup cluster connected by rubber milk and vacuum tubes to a special bucket with a vacuum-tight lid. The lid also carries the pulsator and a rubber tube leading to the vacuum cock at the head of the standing. Milk from the teat-cups is intercepted in the bucket, which is emptied after each cow is milked.

Of recent years "in-can" milking has been introduced, the milk from each teat-cup assembly being received direct into a can "chum". For use in cow-sheds the apparatus may be mounted on a trolley or suspended from a monorail running the length of the shed. Stall cocks may be retained or a vacuum pump may be carried with the rest of the equipment. The system is readily applied to open bails and milking parlours. In all cases the can is suspended from a balance so that the yield from each animal may be determined.

With all types of milking machine the rate of milking increases with greater reduction in pressure, but a vacuum greater than 15 in. of mercury is generally considered liable to result in injury to the animal. It is, therefore, essential that all machines be fitted with accurate vacuum controllers and gauges, and that these components be maintained in perfect condition. The permanent vacuum line should be fitted near the pump with an interceptor trap for collecting condensate and milk in the case of a split liner, and also to facilitate occasional washing. To prevent regurgitation the vacuum line should be laid with a fall and fitted with drain cocks at low points, and, in addition, the vacuum cocks should enter the line at the top.

Milk Strainers : These are usually fitted to the vessel which receives the milk as it enters the dairy. The body of the strainer is fitted with a perforated holder which supports an easily replaceable filter disc, usually composed of cotton wool mounted between layers of fabric, which is discarded after a single milking. Since all milk is filtered immediately before bottling, and there is no evidence that straining on the farm results in improved keeping properties, the practice of straining seems to serve no other purpose than deceiving the dairyman into thinking that the milk has been produced under cleaner conditions than in fact it has.

Milk Coolers : Corrugated surface coolers using water as the cooling medium are extensively used for cooling milk on the farm. The method is simple and cheap but in future, as more emphasis is placed on cooling, it is likely that many installations will be found to lack a sufficiently cold water-supply. Deep bores and wells may be relied onto yield water at temperatures ranging from 50° to 55° F at all times of the year, and such supplies enable milk to be cooled to 60° F. or lower provided always that the water is pumped direct from the well to the cooler and net allowed to become warm in a tank. Many piped-water supplies are, however, much above 60° F. in summer, and it has often happened in the past that little foresight or ingenuity lias been exercised in arranging for the coolest possible water supply for milk-cooling.

Mechanical refrigerating equipment is sometimes used in spite of the heavy capital outlay and the relatively high maintenance and running costs that are involved. The chief application of refrigeration for milk-cooling would seem to be where nothing can be done to obtain an adequate water supply. With such equipment surface coolers may be operated with chilled water, or less satisfactorily with brine or by direct expansion of the refrigerant. It is probably better to do spense altogether with the surface cooler with its attendant problems of cleaning and sterilizer and use the outside of the can as the heat-transfer surface. This may be done by dumping the cans into an insulated tank containing chilled water. The difficulty of lisfting the can of milk may be avoided by spraying the chilled water on to their outsides.

Steam and Hot Water Supply : Although chemical sterilization with hypochlorite solution may be successfully applied to farm dairy equipment, steam sterilization is at present usually considered the method of choice. A supply of hot water for washing is generally provided from the steam raiser, which may be fired with solid or liquid fuel, or may be heated by electricity. The quantity of steam required varies greatly according to the layout of equipment and the care with which the steam is used, but an evaporative capacity of 20 to 50 lb. per hour should be sufficient for medium-sized instal-lations using milking machines.

Since farm boilers are used only intermittently, and because of the large amount of heat left in the boiler and contents at the end of the operation, their efficiency is low. Electric steam raisers may successfully compete with the oil- and coal-fired boilers because their steady heat input allows of the use of smaller water volumes and consequently of lighter shells. Electric equipment is also clean in operation and may, with very little difficulty, be made fully automatic.

Sterilizing Chests : Some form of enclosed space for holding utensils during steaming is required. This may take the form of a sheet-metal chest or small sterilizing room. (All sterilizing chests must be equipped with thermometers.) Where electricity is avail-able, self-contained electric sterilizing chests are popular, and a chest, equipped with a cast-iron heat-storage block, heated with a small electric element, is available. It has the advantage that steam is available at a moment's notice. A suitable arrangement of a milk room designed for bucket or hand milking and using a surface cooler and steam sterilization is illustrated in Fig. 10.

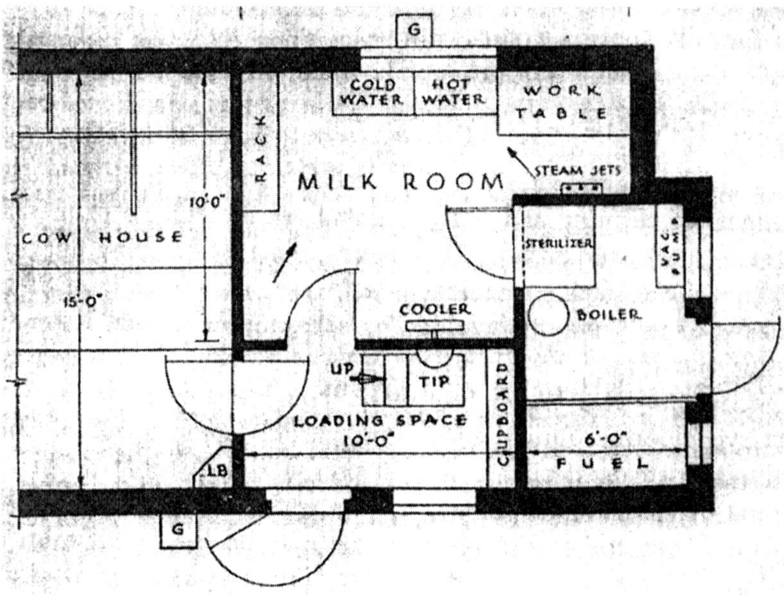

FIG. 10 : Arrangement of Milk Room

Wet Storage : One of the most troublesome pieces of equipment to clean and sterilize is the milking machine teat-cup cluster. Racks are available to hold the teat-cups with the open end upwards and the milk tube at the same level. The whole of the interior of the assembly may then be filled with a cleaning and sterilizing solution. The method originated in the U.S.A., where 0.5 per cent, caustic soda was found satisfactory. Not all makes of machine have clusters that are completely liquid-tight.

MILK SECRETION AND MILKING

In the first-calf heifer, and also in the cow which has had a normal dry period, milk secretion begins sometimes before parturition. In some cases the udder becomes congested and painful several days before calving takes place, in which case it is customary to withdraw some colostrum. If, however, this is done repeatedly over a period of a week or more, the calf will be deprived of colostrum which, as mentioned earlier is important in relation to the calf's resistance to infection.

In the lactating animal milk secretion goes on continuously, causing a build-up of pressure in the udder. Pressure increases steadily for about nine hours, and afterwards at a diminishing rate. If the milk is not withdrawn, secretion is inhibited and milk is re-absorbed. The milk as it is secreted collects first in the alveoli and the minute ducts leading from them, and later in the milk cistern and teat sinus. At this stage the greater part of the udder might be likened to a milk-saturated sponge, with most of the liquid so tightly held that it will not drain by gravity alone; indeed, a tube passed through the milk orifice will drain only the large sinuses. Some milk may be squeezed out of the secreting cavities and tubules by massaging the udder.

Under natural conditions the nuzzling of the bag by the calf stimulates the cow's pituitary gland to release the hormone oxytocin. This is carried to the udder by the blood stream and causes contraction of the alveoli and ducts so that the cow 'lets down' her milk. The effect of the hormone is transitory, and after a period of some seven or eight minutes the milk pressure relaxes so much that even the strong suction exerted by the calf cannot draw the milk from the fine vesicles. Some time must elapse before a fresh stimulus can reproduce the reaction. With the more primitive breeds of cattle the presence of the calf is necessary to produce the response described above, but most cows of the improved dairy breeds, after a few days' experience of hand or machine milking, "let down" completely in response to the handling of the teats and even to a gentle touch on the udder. There are, however, considerable individual differences between "easy" and "hard" milkers. These differences depend in part on the size of the teat orifice. Some cows can be milked out by machine in three or four minutes, whereas others require as long as eight. Moreover, excitement, fear, or pain causes the secretion of adrenal in into the blood stream and this counteracts the effect of the oxytocin

This explains the importance of quiet and gentle handling of cows at milking time.

The pressure of milk in the udder reaches its maximum about one minute after the appropriate stimulus has been applied; so that the timing of the commencement of milking is important. Moreover, since the response lasts for a relatively short period, milking should be done quickly.

The massaging of the udder with a hot moist cloth has been advocated as the best "signal" that milking is about to commence, but it does not seem that this particular process is necessary. Many cows react to other stimuli, including even the noise produced by handling the milking equipment. The important matter seems to be to follow an accurately timed routine of operations upto the point when milking is begun, and to start the operation about a minute after "let down" has been induced.

In hand milking, the teat is first grasped at its junction with the udder, between the forefinger and thumb, so as to prevent the milk from being pushed back into the milk cistern; pressure on the teat is then extended downwards by closing the other fingers, in succession, against the palm, so that the milk is driven out through the teat canal. In the final operation of "stripping," the forefinger and thumb are drawn from the base of the teat downwards along its whole length.

The action of the milking machine, which is different in that both suction and pressure are employed, has already been described. The chief point in machine milking is to prevent the teat cups crawling upwards towards the end of the operation As the milk is withdrawn the udder tissue becomes flaccid and the suction has the effect of drawing the cups upwards until they may press upon and damage the delicate membrane between the milk cistern and the teat, which should open when suction is applied, allowing the milk to flow freely from the milk cisteni into the teat. This difficulty was formerly met by removing the machine before milking was complete and by subsequent hand stripping; but equally good results can be obtained by the machine if the appropriate technique is employed. This is to pull gently on the cups with one hand while massaging the bag with the other. In any case it is very harmful to leave the teat cups on the teats after the udder is empty; damage caused to the tissues through this malpractice is an important pre-disposing cause of mastitis.

Where a regular milking “drill” is constantly observed for a period, a majority of cows can be milked out in a period of three to five minutes. A considerable economy of time is to be gained by grouping the slow milkers together in the shed.

Milking need not be done more than twice a day, except in cases of heavy milkers. A change from three milkings to two will generally produce a diminution of the yield, but the 2 or 3-gallon cow soon adapts herself to the new routine. Heavy milkers have been found to yield from 10 to 20 per cent, more milk when milked thrice daily. The intervals between milking should be as nearly equal as possible, because otherwise the butter-fat content of the milk, after the longer period, will always be low. With milkings at 6 A.M. and 3 P.M., a somewhat common arrangement, it has been shown that the morning milk may be quite 0.5 per cent, lower in butter-that than the evening milk, and hence, may sometimes tall below the legal minimum standard of 3 per cent. Milking should be done quickly and thoroughly. A good hand milker may be expected to milk from seven to ten average cows in an hour and a half, the time ordinarily allotted for the work.

Cows that are still milking at six weeks before their next calving time should then be put dry. It will often suffice merely to stop milking, when secretion will cease and the milk will be re-absorbed. Sometimes partial withdrawal of the milk is necessary, but this should be avoided unless the animal is in pain.

BEEF PRODUCTION

The type of animal best suited to the production of beef has been sufficiently described in the last chapter. There can be no question that the breeder, if beef is his sole object, should aim at the best possible beef type. But the feeder, if he is nothing but a feeder, may find that the finest commercial type of “store” commands a more or less fancy price, and that he can obtain a larger profit from a somewhat rougher animal that can be bought cheaper. The value of a young calf, as material for beef production, depends largely on its breeding. Indications of beef “blood” - a white face or a black or blue-grey body colourare generally a better indication of beef value than size or conformation. Animals with indications of dairy breeding Channel Island or Ayrshire colours and markings-will rarely make profitable beef animals even if they appear to have good conformation.

Calves at six to twelve months of age should show evidence of reasonably good nutrition since inadequate feeding in calf hood usually does permanent harm. Signs of former starvation are small size, a long narrow face, a thin back, and a pot belly.

On the other hand, "hard wintering" of older cattle does little permanent harm, in particular, cattle that have been out-wintered and fed rather poorly commonly make much better live-weight gains, on grass, then, such as have been housed and well-fed. The latter often lose weight on the spring grass.

The slaughter age to be aimed at depends on breed. The hardy mountain breeds do not respond to forcing, and the dual-purpose types require more time than the early maturing beef breeds.

Angus, Herefords, and Shorthorns of beef type, if they are well-fed as calves, and if they are carried on without a " store" period, can be marketed either as "baby beef at fifteen to eighteen months or as young beef at eighteen to twenty-four months. This system is well-adapted to the better sort of mixed farms where breeding and feeding are combined. On the other hand, feeders who buy in stores and who work on a system of rapid turnover prefer cattle of some age because they fatten more quickly. Two and a half years is about the average age for cattle to be put up to feed, and some of the later-maturing breeds are kept till they are three years old or over. Old and heavy cattle, however, sell in a free market at a considerable discount as compared with young, well-finished animals. Under the system of payment according to grade, which operated fiom 1939 till 1954, the price per hundred weight was based mainly on the estimated dressing percentage, and the premium formerly commanded by light-weight young cattle was considerably reduced. With the' restoration of free markets in 1954, a considerable premium for cattle of moderate size and good quality has again become established.

The general scheme of feeding must be adapted to the prevailing economic conditions. If the fattening process is likely, in itself, to be a profitable operation-i.e. if there is any considerable margin between the value per hundredweight of lean and fat animals, and if feeding stuffs are cheap-the general rule will apply that the more rapid the process of fattening the higher will be the profit. Obviously, if one animal is fattened in four months and another, on a lighter ration, to the same degree in six, the latter is debited with the cost of an additional two months' maintenance, without any compensating advantage. In the past half-century, however, the conditions stipulated have rarely existed, and the winter fattening of cattle has come to be regarded as essentially a means for the disposal and conversion into manure of the bulky and non-marketable by-products of the arable or mixed farm. Hence, it has usually been good policy to feed a maximum of such by-products, and to confine the use of purchased concentrates to the barest minimum. Generally, some such concentrates will be necessary, if the animal is to obtain sufficient nutrients to enable it to make reasonable progress, and also, in some cases, to correct the protein deficiency of roots and straw.

As regards winter feeding, the ration will depend mainly on the particular home-grown feeding stuffs that are available. On arable farms in Scotland the chief constituents will generally be roots and straw, with a moderate allowance of seeds hay in the latter part of the fattening period. In the eastern countries of England hay is the usual fodder, and sugar-beet by-products largely replace roots. Silage is a good substitute for roots and hay where roots cannot be grown economically. Good silage, made from any of the ordinary mixtures of, say, oats, beans, peas, and vetches, has nearly twice the energy value, weight for weight, of swedes, and more than twice the protein equivalent, so that expensive, highly nitrogenous concentrates become unnecessary. On the other hand, its fibre-content is considerably higher, so that its use leads to a reduction in the amount of straw, etc., that the animal will consume. In semi-arable districts meadow hay will generally be available in quantity and, in pastoral regions, it may form the basis of the ration. In the colder and later arable districts, where grain is often of poor market quality, oats and barley are largely used.

During the progress of fattening, as the animal's appetite for bulky food declines, the allowance of concentrated food must be increased. Wood's standard for a fattening steer allows 12 lb. of starch equivalent and It lb. of protein equivalent per 1000 lb. live weight. On such a ration a 1000-lb. animal should add nearly 2 lb. daily to its live-weight, more in the earlier stages when the live-weight increase consists partly of water and protein, less in the latter when the gain is chiefly fat. Good cattle will readily make such an average increase, as between ordinary store and ordinary fat condition. On more highly concentrated rations 2 ¼ to 2 ½ lb. per day may be obtained. In practice, as has just been indicated, it may often be wise to rest content with a daily increase of It to 1 1/2 lb. using a less nutritious ration. In countries where cheap concentrates are available *(e.g.* maize in the corn belts of North and South America) rations with as high a starch equivalent as 18 or 20 lb. are sometimes fed, and the gain may reach 3 or even 3 ½ lb. per day for short fattening periods. Such rations are often imperfectly digested, and pigs must be kept with the cattle to consume the residue.

The following are examples of winter rations for arable and semi-arable conditions, and for the beginning, middle, and end of the fattening period, respectively. The cattle are assumed to weigh 8½ cwt. at the beginning, and to make an average live-weight increase of 2 lb. per day over a feeding period of five months. The fat weight, would, thus, be about 11 cwt.

Arabic Conditions

Beginning	**Lb.**	**Middle**	**Lb.**	**End**	**Lb.**
Turnips or sugar-beet tops	80.0	Swedes[1]	70.0	Swedes	60.0
Oat straw	14.0	Oats straw	12.0	Oat straw	6.0
Palm-kernel cake	4.0	Palm-kernel cake	2.0	Seeds hay	6.0
Oats	2.5	Linsed cake	2.0	Linsed cake	3.0
		Maize meal	3.0	Barley meal	5.0
Dry matter	24.0	Dry matter	24.0	Dry matter	25.0
Starch equivalent	11.0	Starch equivalent	12.5	Starch equivalent	13.5
Protein equivalent	1.4	Protein equivalent	1.6	Protein equivalent	1.8

Semi-arable Conditions

Beginning	**Lb.**	**Middle**	**Lb.**	**End**	**Lb.**
Turnips	42.0	Swedes	42.0	Mangolds	42.0
Oats straw	6.0	Meadow hay	16.0	Meadow hay	14.0
Meadow hay	14.0	Barley meal	4.0	Barley meal	6.0
Barley meal	3.0	Palm-kernel cake	3.0	Linseed cake	3.0
palm-kernel cake	2.0				
Dry matter	25.0	Dry matter	25.0	Dry matter	25.0
Starch equivalent	11.0	Starch equivalent	13.0	Starch equivalent	13.5
Protein equivalent	1.4	Protein equivalent	1.7	Protein equivalent	1.8

Much thought was being given, at the time of writing, to reduce the costs of winter feeding. Many cost accounts, drawn up on conventional methods, have indicated negligible profits or, [1] Swedes or mangolds may

[1] Swedes or mangolds may be largely replaced by dried sugar-beet pulp.

be largely replaced by dried sugar-beet pulp. actual losses; but it should be noted that the value of dung, under certain conditions, is in fact much higher than that allotted on the basis of the residual manurial values of the food and litter. This is especially true on the lighter sorts of land where high value crops, such as potatoes and many vegetables, are grown.

The high costs arise mainly from the high charge of labour requirements of growing fodder roots, of tending housed cattle, and of carting and spreading the manure.

One approach is to shorten the period of house feeding by running the cattle on foggage. beet tops, and kale after the end of the normal grazing season: another is to use silage for a substantial part of the ration; satisfactory result have been obtained on rations composed mainly of silage and cereal foods.

"Baby beeves" *i.e.* such as arc being fattened while still in rapid growth-require a more concentrated type of ration. Straw cannot be successfully used and even hay must be so limited that not more than a third of the dry matter is supplied "in this form. In all cases the choice of particular concentrates should be determined by ruling market prices.

Apart from questions of bulk, energy value, and protein content, the palatableness of the ration may require special consideration. Some feeding stuffs, such as palm-kernel cake, are not readily eaten at first but are consumed freely enough when the animals have had a week or two to become accustomed to them. An occasional animal will refuse any and every sort of cake. Sprinkling with molasses water, or admixture with any of the commercial condiments, will often be helpful. Again, stores from grassland districts take some time to learn to break roots, and an occasional animal gives up the problem altogether. If it is inconvenient to slice the roots for the whole group, such individuals should be transferred, if possible, to individual boxes and given special feeding. Regarding the preparation of the food generally, roots must be sliced for cattle that are in process of changing the first pair of incisor teeth (say eighteen to twenty-four months old). Otherwise it is doubtful whether any special preparation of the food is profitable. Many experienced feeders give the roots whole, the fodder long, and the cake in the ordinary form of nuts or cubes. On the other hand, some feeders chaff a proportion of the fodder, slice the roots, and nix the two together, with or without the addition of meal. This doubtless decreases the work of mastication and enables the animal more quickly to finish its meal and return to rest, but it involves a large amount of additional labour, and mere is no satisfactory evidence that the advantage is sufficient to leave a profit. Only if straw is not being consumed in the quantity desired, owing, for example, to its tough and dry condition, it may be chaffed and either mixed with pulped roots or moistened, and left for a day to ferment slightly before being fed. The concentrate should not be added to the mixture in such cases, as the fermentation causes loss of nutrients without any compensating advantage. Heavy root rations are always liable to cause scouring, particularly in early autumn; materials like soyabean, earth-nut, and linseed cakes, which have also a laxative tendency, should not be fed with them; undecorticated cotton cake, on the other hand, is an excellent corrective.

The particular details in the arrangement of the ration are of less consequence than strict adherence to whatever scheme is adopted. In practice straw is generally given *ad libitum,* and the racks are filled morning and evening, or oftener, if necessary. The roots should be given in two feeds or, if the ration is over 80 lb., in three; the concentrate should also be given in two meals, if the amount exceeds 4 or 5 lb. Many feeders believe in allowing a period of two or three hours of entire quietness in the middle of the day. Cattle receiving very large rations of roots will not require drinking water, but, ordinarily, water should be laid onto the yards. The total consumption of water (including any in the food) is generally about 1 gal. per cwt. live weight per day, or approximately four times the weight of the dry matter eaten.

Regarding housing, fatting cattle may be kept in single or double boxes, or may be run in lots of upto twelve or twenty in covered or partially open yards; or, again, they may be tied up in stalls, as are dairy cows (a common practice in the north). Single boxes doubtless give the best results, but the method is expensive both in labour and in housing space. Stalls are economical of space, 50 sq. ft. being sufficient floor space tor a beast against nearly 100 in covered courts. Stalls are likewise economical of litter, 7 lb. being a sufficient daily allowance, while about twice this amount is needed in yards. The labour is, however, increased as compared with other methods owing to the necessity of individual feeding, daily removal of manure, and occasional grooming. The conservation of the manure will be less satisfactory than where it is kept firmly trampled in yards. Partially open courts are wasteful of litter and

PLATE 10

A. ENGLISH LEICESTER RAM

B. BORDER LEICESTER EWE

PLATE 11

A. HALF-BRED SHEARLING RAM

B. WENSSLEYDALE RAM

manure, but are otherwise satisfactory except in very exposed situations. Cattle running together in lots are always liable to disturbance from restless or pugnacious individuals, and occasional beasts are prevented from getting their share of food. Lots usually settle down best, if they have been grazed together for a month or two before being housed. In any case, horned cattle should have the tips of the horns removed before they are housed in lots, otherwise there is endless disturbance and a consequent loss of weight. A limited amount of stall or box accommodation is always necessary in order that bullying or fright-ened or sickly animals may be segregated.

Summer fatting of cattle may be carried out on the best sort of permanent grass without additional feeding, the land carrying a full-sized bullock to the acre and providing, in addition, some autumn and winter grazing for sheep. Well-managed rotation grass on good arable land will often do as well as this. Very careful adjustment of the number of cattle to the growth of the grass is necessary if the cattle are to make full progress without cake. On the one hand, if the grass gets overgrown the fibre-content rises and the digestibility falls; on the other hand, if the pasture gets too bare the cattle fail to get a full ration in the time (about eight hours in the twenty-four) that they spend in grazing. In the first case, the mowing machine should be used to keep the grasses under control. In the second, the farmer must resort to cake feeding. The type of stock should also be adapted to the quality of the pasture. In the chief grass-fattening districts of the Midlands the best fields are stocked with steers, the slightly poorer ones with heifers, and those of still poorer quality with barren and other cows culled from dairy herds. Of average feeding pasture about 11 acres should be allowed per head, and some concentrate will often be needed during the latter part of the season. The concentrate used should be one of high energy value and low protein equivalent. Maize germ meal, maize meal, barley meal, and rice meal, made up for convenience in cube form, give satisfactory results. Live-weight gain is, of course, made at a much lower cost on summer pastures than on winter rations, but the grazier must buy his stores in a dear market (April and May) and usually sells a large proportion of his fat stock at the cheapest season-August to November.

The degree of fatness to which cattle should be brought before marketing depends on market conditions. It must be realised that during the last stage of feeding-that which makes the difference between "good" and "prime" condition-the liveweight increase is obtained at a very high cost. On the other hand, when prime cattle are scarce, the extra cost may be more than repaid in the increased value per hundred weight. The most serious error to make cattle fatter than the butcher desires, for in this case a low price is obtained for an article with a high cost of production. Ordinarily fat cattle will dress about 57 per cent, of their live-weight, or, say, 8 lb. to the stone. Prime cattle may reach 62 or 63 per cent., and exceptional animals in very high condition from 65 to 70. Very lean animals may yield under 50 per cent, of carcass. The largest items in the "sinkage" are the contents of the alimentary canal (15 per cent.) and the hide (7 per cent.). Dressing percentage will obviously vary considerably with the quantity of food contained the stomach and intestines; cattle fasted and thirsty may weigh 70 or 80 lb. less than when full.

CARE AND MANAGEMENT OF BULLS

Bull calves intended for breeding should be kept in growing and thriving condition from the start. Those of the beef breeds are normally nursed by their own dams, though occasionally they are changed over to foster-mothers which yield relatively large amounts of milk. Dairy bull calves are often suckled by cows which, because they are "hard milkers" or for other reasons, are troublesome in the dairy herd. Alternatively they may be reared on the bucket. After the calfhood stage, bull calves should continue to be well-nourished but not necessarily made fat They should be housed in airy boxes, kept practically at outdoor temperature, and should have access to an open yard. Exercise is essential, and this may be provided by a rim on pasture for an hour or two daily, or alternatively a walk in halter. Subsequent management is always easier, if the young animals have been regularly haltered and led about. The feet must be kept 2in good shape by occasional trimming, and the coat should be kept clean by grooming and an occasional washing. A nose-ring is inserted before the animal is strong enough to cause trouble-say at the age often or twelve months. Either a rope or a pole should be attached (by means of a spring hook) to the nose-ring when the bull is led out, but control should normally be exercised by the halter, and the nose-ring kept in reserve in case of trouble. If further restraint is necessary, as it occasionally is in the case of older dairy bulls, blinkers may be put over the eyes, so adjusted that the animal can see only the ground in front of it.

Bulls must be inspected for licensing before they are ten months old, and it is obviously important tot they should be presented in good condition. Breeders not infrequently complain that bulls are rejected for no other reason than lack of condition, but it is obviously very difficult for an inspector to satisfy himself that the condition of a lean and somewhat ragged specimen is due entirely to its restricted diet. In the case of beef breeds-and to a less extent in that of dual-purpose types-capacity to fatten, and to fatten in the right way, is of fundamental importance, and it is impossible to say with certainty whether a lean animal possesses the capacity in the desired degree.

It is of obvious importance that bulls should be retained until the merits of their progeny can be assessed i.e. in the case of a beef breed for about two years after the animal has been put into service, and in that of a dairy sire fully four years after use has begun. In fact very few dairy bulls are kept so long, chiefly because a large proportion become progressively more difficult to manage as they get older. Another factor is that if an old bull is kept on after his daughters reach breeding age, a second sire is required to mate with these, and the keep of two males is a very heavy charge in the case of the average herd of perhaps fifteen cows. There is no question that the solution of the bull problem, for the smaller sort of dairy farm, is artificial insemination; but large-scale breeders, and especially owners of pedigree herds, will probably continue to keep their own bulls even if they make some use of the artificial insemination service.

The bull box should be light and airy, and should preferably have access to a large open yard. Failing this yard, opportunity for exercise may be provided by means of an overhead wire with a runner to which the bull's nose-ring can be attached by means of a chain, or by one or other of a variety of devices. Where there is a yard outside the box the door communicating between the two may be of the portcullis type and be operated from the feeding passage of the box, or from outside, by means of a wire rope; this arrangement enables the bull to be restricted to the yard while his box is being cleaned and littered, or *vice versa.* This is easy to arrange that the bull may be secured, by his nosering, from the feeding passage, the animal being brought within reach by the offer of food. Again, a service pen may be so placed and devised that the bull may serve a cow and be put back into his yard without being handled or being given any opportunity to attack his attendant. These various precautions may seem to be unnecessary in the case of good-tempered animals, but a large proportion of bulls are subject to occasional and unpredictable fits of viciousness.

Apart from the important matter of exercise, the bull must be kept fit by adequate nourishment and yet not allowed to become fat. A suitable ration for a mature animal of average size is 20 to 25 lb. of hay, with a few roots in winter and some green stuff in summer, and perhaps 2 lb. of the concentrate mixture that is fed to the milking cows.

If the bull-yard is laid with well-roughened concrete the animals feet may not require much attention, otherwise the hoofs must be trimmed back, when they get too long. When such attention is needed, some animals may be tied up and handled as horses are when they are shod, but many must be put in stocks, or alternatively, be hobbled with special tackle and thrown in order that the operation may be carried out.

CHAPTER V

SHEEP AND GOATS

SHEEP and goats, with certain intermediate species, are generally regarded as forming one group of the Cavicornia or hollow-horned ruminants.

The true sheep comprise a considerable number of wild species which fall into two rather distinct natural groups. The one includes the Mention *(Ovis musimon)* of Corsica and Sardinia, together with the closely related species of Cyprus *(O. cypria)* and of Asia Minor *(O. oriental is);* and the Urial *(O. vignei)* of Persia, Northern India, and Tibet. The other group consists of the Argali *(O.ammon)* of Western Siberia and the Altai; Marco Polo's sheep *(O. Pol£);* and the Bighorn of Northern Siberia *(O. nivicola)* and of the Rockies *(O. montana).* The true goats include the Markhor *(Capra falconeri)* of Afghanistan and the common wild goat or Bezoar *(C. aegagrus)* of the Caucasus, Persia, and Crete. The Steinbok and the Tur are included in the genus *Capra,* while the Bharal or Blue Sheep of Tibet, the Barbary Sheep, and the Tahr of Nepal and Kashmir, represent types intermediate between sheep and goats, and are referred to separate genera.

Nothing is known for certain regarding the origin of domesticated sheep. The view most widely held is that the Asiatic Mouflon supplied most of the foundation stock, but the Urial and even the Argali have sometimes been regarded as the ancestors of some domestic types. The older view, to the effect that the wild ancestor must have been quite different from any surviving species, is not now accepted. The common goat is descended from the Bezoar, and the Cashmere goat from the Markhor, while the valuable Angora is generally believed to have been derived from a cross between the two.

Sheep are valued commercially for their mutton and wool, and occasionally also for milk. With some breeds, such as the Merino, wool is the primary product, while British breeds are kept mainly for mutton-wool being a secondary, if still an important, consideration. The points of the mutton sheep correspond rather closely to those of the beef ox, the valuable qualities being a tendency to fatten early in life (early maturity)- a thick covering of lean meat and an even and not wasteful distribution of the fat. The head varies greatly with the breed, but should be of good depth and width, strong of jaw, and broad through the nose; the neck moderately short, and thick at its junction with the body; the body deep, wide, and square; the shoulder top wide, level, and well-covered; the chest deep and wide, and the ribs well sprung or rounded; the shoulder neatly laid and thickly covered, and the region behind the shoulder well filled up; the hind quarters long, broad, and level, with the width well carried out to the rump; the gigot or leg of mutton thickly fleshed, and the flesh well-carried down, both inside and out. The back should be wide and level throughout, thickly covered with firm, muscular flesh; the legs moderately short and set well apart, the bone neither too coarse nor too fine, but clean and hard. The thickness of the tail is a useful measure of condition or fatness. The carriage should be stylish and springy, indicating activity and robust health.

Sheep's wool is commercially by much the most important fibre of its kind, although other wools and hairs-for example, mohair, cashmere, alpaca, and camel's hair-are all used as textiles. Wool is used in the manufacture of a great variety of articles. The chief are, of course, articles of clothing for the inhabitants of the colder regions of the world, but carpets, tapestries, and blankets absorb a considerable quantity of certain classes. The characteristics of the wool fibre on which its value mainly depends are as follows :

1. *Length : This* varies from under 1 in. to over 2 ft., the coarser wools being generally the longer. Other things being equal, the longer wools are the more valuable.
2. *Uniformity in Length of the Different Fibers in the Staple :* This is important, because, for certain purposes, the short fibres require to be separated and put to a different use, bringing a lower price.
3. *Strength and Elasticity : Strong,* elastic fibres suffer a minimum of breakage in the process of manufacture and naturally give a strong yarn. Fine fibres are stronger in proportion to their size than coarse. The staple should be free from any weak part or " break," as wool with such a " break" cannot be combed without great wastage. The condition is generally due to the sheep having suffered a severe check during the period of growth of the wool. In Britain, the common causes of a " break" in the staple are a period of

PLATE 12

A. LINCOLN LONG WOOL RAM

B. SOUTH DEVON SHEARLING

A. KENT OR ROMNEY MARSH SHEARLING RAM

B. SOUTHDOWN SHEARLING EWES

PLATE 14

A. SHROPSHIRE SHEARLING RAM

B. SUFFOLK RAM LAMB

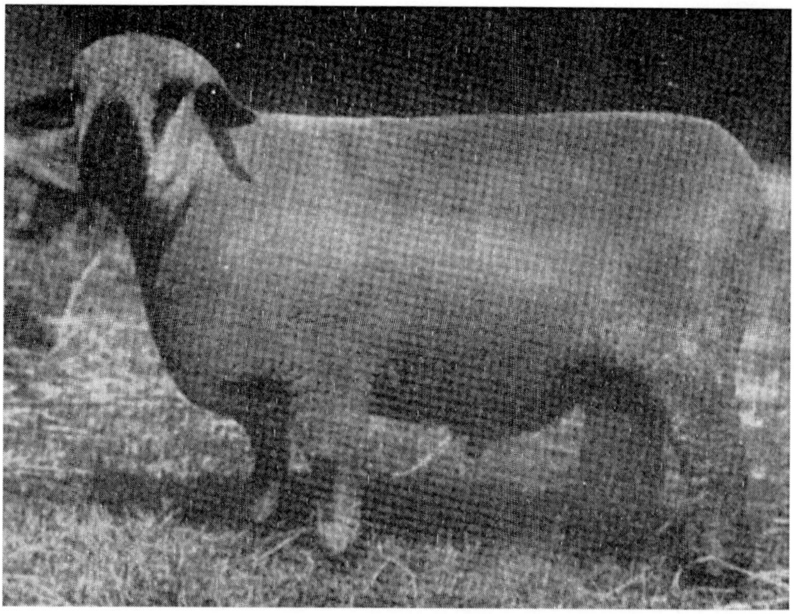

A. HAMPSHIRE DOWN SHEARLING RAM

B. OXFORD DOWN SHEARLING RAM

undernutrition or partial starvation, which may affect the whole flock (particularly in the case of hill sheep), and illnesses of various sorts. The weaker part of the fibre is that produced at the end of the period of malnutrition e.g. just before the commencement of fresh growth on a hill grazing. In extreme cases the sheep casts part or the whole of its fleece.

4. *Fineness : The* mean diameter of the fibres may vary from about $^1/_{1500}$ in. in the finer Merino wools to $^1/_{1600}$ in. or more in the coarser long-wool fleeces. The smaller the individual fibres the finer the yarn that can be spun from them, and in yarns of the same weight the finer wool gives a softer and fuller character, which in turn gives the finished cloth a better appearance and "handle." Softness is due chiefly, but not entirely, to fineness of the individual fibres.
5. *Crimp :* The aptitude of the different fibres to interlock depends on the fineness and extent of the waviness or "crimp." Fine Merino will often show more than twenty waves to the inch, while most Longwools give only three or four. Crimp also gives springiness or elasticity to the finished cloth. Felting further depends on the number of epidermal scales and on the extent to which their upper edges project. Fibres with a smooth surface, like those of the Lincoln or Wensleydale sheep, are naturally slippery. On the contrary, Merino or Down wool (which lias the appearance under the microscope of a series of irregular flowerpots set one inside the other) felts readily. The wool of a healthy sheep will rarely felt or " cot" on the sheep's back because the fibres lie parallel, with the serrations all pointing in one direction, and because of the coating of oily secretion "Cots" or felted fleeces do, however, occur, and have a low value.
6. *Colour : A* clean white or pale lemon-coloured wool can be dyed to any desired shade, whereas coloured wools, or those containing a proportion of black or brown fibres, have a restricted use. Hence dark fibres are objectionable, and every endeavour should be made by the breeder to eliminate them by selection. Certain natural-coloured wools-e.g. "moorit" (brown) Shetland -have a higher value, for special purposes, than white wools.
7. Lustre : Wools with large, flat, epidermal scales have a bright shiny appearance, whereas if the scales are small and the surface rough, the wool is dull in appearance. High lustre and felting qualities are obviously incompatible.

8. *Behaviour to Dyes :* True wool absorbs dye differently from what is called "kemp," which is a coarse and generally brittle fibre with a characteristic dead-white appearance. It does not absorb dye in the ordinary dyeing process, and hence causes unsightly marks in the cloth. Here, again, the breeder should make every effort to get rid of the fault by careful selection of his breeding sheep. Kemp is commonest in the wool of mountain breeds, and generally increases with the age of the animal.

9. *Uniformity of the Different Portions of the* fleece : Previous to its manufacture, fleeces are "classed" according to their general quality, and each fleece is then "sorted" *i.e.* it is pulled to pieces and these are placed in separate lots according to quality. Classing is done either by the producer, the merchant, or in many cases by woolgrowers' co-operative societies. Sorting is normally carried out by the manufacturer who, however, may re-sell the lots that do not meet his particular requirements. Sometimes, as in the best Merino wools, the fleece is practically uniform throughout. In most sheep the quality varies according to the part of the body from which the wool is taken. That from the shoulder is generally the most valuable, being deep in staple, uniform, and fine. The neck wool is often finer, but is much less uniform; that from the breech, on the other hand, is both coarse and irregular. The wool-sorter divides the fleece into different " counts," the count representing the number of hanks of yarn (each 560 yds. long, according to the Bradford system), tot could be spun from 1 Lb. The count, as usually applied, is a measure of fineness, although in actual spinning the length of yam that can be got from the pound of wool varies with the length as well as with the diameter of the fibres. A good average Merino fleece might give wool of 64 quality from the back part and 70'S from the front. A typical Longwool *(e.g.* Lincoln) fleece sorted into its separate counts is shown in Fig. 11. Skirtings, cots, brand marks, and soiled locks are removed before or during the process of sorting.

Wool is manufactured by either of two methods. In the worsted process the raw material, after scouring, is combed so that the short fibres are removed. The latter are sold as " noil" to the woollen manufacturer, while the long fibres, arranged parallel to each other in a sort of rope, form "wool tops." The tops are spun into a rather firm, hard thread, the cloth made from which has a distinct thread structure and a bright appearance. The important qualities in worsted wools are length, strength, uniformity, trueness, elasticity, lustre, and fineness. High felting qualities are generally not required. In the second or *woollen* manufacturing process the short fibres are not removed, length being unnecessary. The material is carded that is, it is teased out into a thin web in which the fibres run in all directions. The carded "silver", or rope, is condensed and spun without further treatment. The cloth is more or less felted by moisture and pressure, so that the fibres are interlocked in all directions and the thread structure becomes indistinct. The important qualities for this purpose are fineness, softness, and feltig properties. A break in the staple is an unimportant defect, and lustre is not wanted, since the haphazard arrangement of the fibres would, in any case, conceal any natural lustre of the wool.

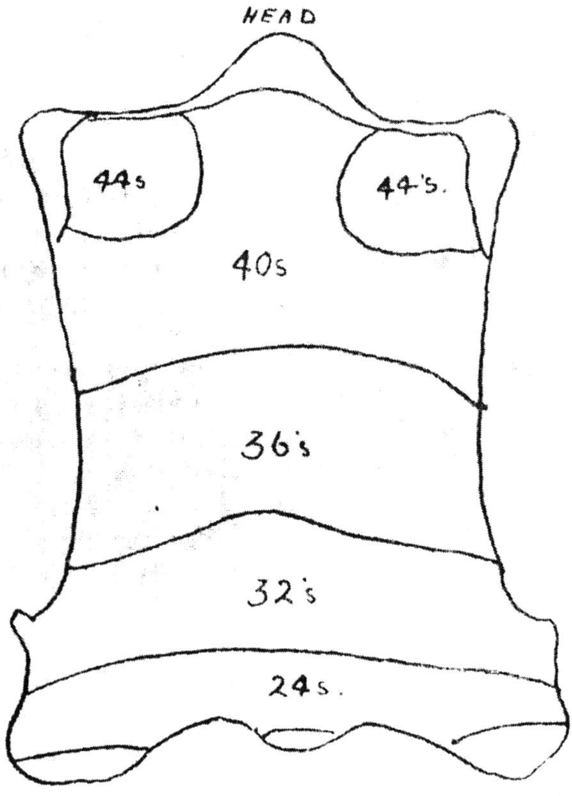

FIG. 11 : A Longwool Fleece, shwoing Sortings (Jones)

The value of natural or "greasy" wool will, of course, vary with the proportion of fatty and sweaty matter that it contains, as well as with its moisture content. The "yield" of scoured wool, after the removal of the "yolk", may be has high as 80 or 85 per cent. in the Longwools, and may fall to 60 or even as low as 30 per cent, in some Merinos. The general run of British wools from unwashed sheep may give an average yield of about 70 per cent. Vegetable impurities, such, for example, as straw, burrs, grass seeds, or fibres *e.g.* from jute packs or hemp stringare all very objectionable in wool, since they do not absorb dyes in the same manner. They are often difficult to separate, and their presence in large quantities may render it necessary for the wool to be carbonized {treated with acid and heat) in order to destroy them. Hence, every endeavour should be made to keep them out. As extraneous matter is readily picked up during shearing, this process should be carried out on a clean floor or on a tarpaulin. Sheep are commonly branded after clipping. Tar should not be used for this purpose, but one of the materials officially approved by the wool trade. Sheep are frequently coloured "bloom dipped" for sale purposes, and some of the materials used for this purpose permanently discolour the wool; such should be avoided. Shearing should never be done when the fleece is damp, otherwise the wool will deteriorate in storage. The fleeces must be stored under dry conditions.

From the farmer's point of view the weight of the fleece is obviously an important consideration. This depends on the size of the sheep and on the length and density of the fleece. Density is judged by grasping firmly with the hand. Single fleeces of modern breeds vary from 2 or 3 upto 15 lb., and exceptionally up to about 30 Lb.

The wool breeds of sheep, as distinct from those in which mutton is the prime consideration, are all derived from the Merino of Spain. The Spanish sheep is now of little account but its descendants have covered a great proportion of the globe, and many distinct types have been evolved, notably in Australia, the United States, Germany, and France. Of these the American "C" type and Delaine Merino, as well as the French Rambouillet, are relatively large, robust, smooth-bodied sheep, with wool of good length (3 in.) and moderately fine quality, and capable of producing lambs of good quality for slaughter. At the other extreme is the American "A" type, small and of very poor mutton form, with a very wrinkled skin and a heavy clip of very fine but short and very greasy wool. Merinos are best adapted to dry and at least moderately warm climates, and are a profitable type of stock for land at a great distance from centres of population.

BRITISH BREEDS

Omitting a few of minor importance and the Roscommon and Galway breeds of Ireland, there are thirty-two British breeds of sheep, falling into three main groups :

A) **Longwools** embrace the Leicester, Border Leicester, Wensleydale, Lincoln, Cotswold, Devon Longwool, South Devon, and Kent or Romney Marsh. These are, broadly speaking, the largest breeds They are white-faced (except the Wensleydale), hornless, and have a fleece of long lustre wool which is used for worsted manufacture They mature fairly early and fatten readily, but the mutton, at least of the older animals, tends to be coarse grained and poor in flavour, and readily becomes too fat.

B) **Shortwools** include six breeds that are ordinarily spoken otas Downs, and seven others. The Downs are hornless, with darkfaces and legs, and generally a short, close fleece of relatively (inequality which is suitable for hosiery manufacture. Most have a strong flocking instinct and are docile, so that they are easily kept in folds.The mutton is generally of very good quality, with a high proportion of fine-grained lean meat and a fine flavour. Originally associated with the Chalk Downs, they have now spread all over Britain, except the mountains, and are kept under a great variety of conditions. The Down breeds are the Southdown, Shropshire, Suffolk, Hampshire Down, Dorset Down, and Oxford Down. The Oxford isnot a typical member of the group but is intermediate between theDown and Longwool types. The remaining short-wooled breeds arethe Dorset Horn and the Western or Wiltshire Horn (white-facedhorned), the Ryeland and Devon Closewool (white-faced hornless), the Kerry Hill (speckled-faced hornless), the Clun Forest (dark-faced hornless), and the Radnor (brown or tan-faced).

C) **Mountain** Breeds : These have little in common except their generally small size, good quality of mutton, and their ability to live on mountain or moorland grazings. They include the Scotch Blackface, Swaledale, Lonk, Rough Fell, and the Dales-bred, all belonging to the black-faced, coarse-wooled, horned type. The Derbyshire Gritstone is sometimes grouped with these, but is softer wooled and hornless. The others are the Cheviot, North-country Cheviot Mountain, South Wales Herdwick, Exmoor Horn, Dartmoor, and Shetland. The quality of the wool varies widely even within particular breeds. The coarsest is used for carpet-making, the intermediate for Harris and other strong tweeds,.and the finest for hosiery and flannel.

The Leicester (Plate 10, *A)* is the oldest, or at least was the earliest improved of the Longwool breeds. It was evolved by Bakewell between 1755 and 1790 from the pre-existing Old Leicester type. The latter was large and coarse, late maturing, and of poor mutton type, though it bore a good fleece of long combing wool and was hardy and prolific. Bakewell effected a very great improvement in early maturity and in mutton qualities generally, but it is generally acknowledged that he allowed the fleece to deteriorate and the fecundity to decline. Bakewell's type would doubtless carry far too much fat for modern taste, but fat meat was at his time in demand. During and after Bakewell's time the Dishley Leicesters spread all over the country; they gave rise, as an offshoot, to the Border Leicester, and they were used in the improvement of all the other Longwools and of some other breeds as well. With the change of taste in favour of smaller and leaner mutton, the breed, as a commercial sheep in its pure bred form, has declined in popularity, but it is still useful for crossing purposes.

The modern Leicester is a medium-sized Longwool of low-set and blocky conformation The carcass weight of mature ewes is from 100 to 140 lb., rams reaching 200 Lb. The head is wedge-shaped, the nose slightly Roman and rather thin at the bridge; the hair on the face white with a bluish tinge and dark spots, and the lips and nostrils dark. The forehead is covered with wool, the ears are bluish or white, long and thin; the neck is short and thick, and is carried almost horizontally. The fleece is fairly dense, of good length, and is fine, curly, and highly lustrous. The count is about 46/40. Good ewe flocks will clip about 10 lb. per head and tegs 12 to 14 lb. The mutton is of fair average Longwool quality.

The chief centre of the breed is now the Wolds in the East Riding of Yorkshire. Some flocks are still kept pure for ordinary commercial purposes, but there is a growing tendency to cross the Leicester ewe with the Suffolk ram in order to obtain a leaner type of mutton with better flavour; other flocks are maintained in order to produce rams for mating with Mountain, or more generally, with " Masham" (Wensleydale x blackfaced mountain) ewes. Abroad, the breed has been successful in Australia, New Zealand, and Tasmania, and to a less extent in South America. It is also found in Canada and the United States, but there it has been merged to some extent with the Border Leicester.

The Border Leicester (Plate 10, *B)* is descended from Dishley sheep that were brought to the Scottish Border district from about 1800 onwards. Whether or not any Cheviot blood was introduced in the formation of the Border type has been much disputed. English Leicester rams continued to be imported by Border breeders until about 1850, but since that time the types have been kept distinct. By 1869 the two types had so far diverged that the Highland Society decided to institute separate classes for each.

This breed is of similar weight to its relative, but is higher on the leg, with a high and straight underline. The head is long, rather thin through the nose and light behind the ears. The nose is aquiline, the whole head free from wool, and the lips and nostrils black. The ears are large, white with occasional black spots, and carried rather high. The neck is distinctly long, but thick at its junction with the trunk, and the head is carried gaily and high. The breed is exceptionally wide and level of back, and is very easily fattened. The modern tendency is to aim at a rather short and dense fleece. The wool is wavy and lustrous, with an average count of 48/44; 10 lb. is reached as an average clip in the best flocks, and 14 lb. is a common clip for well-bred and intensively fed ram hoggs. The Border Lejcester is fairly hardy and extremely prolific. Flocks kept under good conditions will often produce 170 per cent, of lambs, and the ewes are good milkers. The chief use of the breed is the production of rams for crossing with Cheviot and Scotch Blackface ewes. It is admirably adapted for mating with the smaller and leaner breeds owing to its combination of size, early maturity, and a thin head-which last is an important feature in connection with parturition Border Leicesters are largely bred in Northumberland and Cumberland and throughout all the lowland districts of Scotland, and the breed is common in Northern Ireland. A blue-faced strain with a finer fleece is bred in the vicinity of Hexham and is preferred locally for crossing with the black-faced mountain breeds. The Teeswater is somewhat similar. The Border Leicester has been largely exported, and is probably the most widely known of the Longwools. In Australia, it is the most numerous of all British breeds, and gives excellent results, when crossed with the Merino.

Half-bred (Plate *11, A)* or Border Leicester x Cheviot ewes are the common type of breeding stock on arable farms north of Yorkshire, and are an important commercial type in many grassland districts in the Midlands and elsewhere in England.[1] They are hardy, very prolific (rarely giving less than 150 per cent, of lambs), and excellent mothers, and when crossed again with Oxford, Suffolk, or other Down rams, produce an excellent type of butcher's sheep. The Wensleydale (Plate 11, *B)* was produced originally by crossing the Dishley Leicester with an old, large, white-faced breed known as the Teeswater "Mug." The characteristic blue face and legs were derived from the half-bred Leicester ram "Bluecap," born in 1839. The modern type was fixed about 1860.

The Wensleydale is a large-framed sheep, rather long of leg and long of body. The depth of rib is good, but the back is less broad than that of the Leicester. The head is strong, highly carried on rather a long neck, and the forehead carries a tuft of wool. The fleece is long, similar in quality to that of the Leicester (46/40) but with a higher lustre. It is divided into locks of about pencil size and is open in character. Good ewe flocks average upto 9 Lb. of wool. The face and legs are of a very characteristic deep blue-grey colour. Some strains produce a considerable proportion of black lambs.

The Wensleydale is probably the hardiest of the Longwools, but is notably narrower in build than the Leicester or Lincoln and is distinctly later maturing. The mutton is of good quality with, for a Longwool, a remarkably high proportion of lean meat. The breed is mainly found in the North and West Ridings of Yorkshire and in Lancashire. Its chief use is the production of rams for crossing with the various black-faced mountain breeds, both in the north of England and in the south-west of Scotland. Crosses with Swaledale ewes are known as Mashams (Massams) and those with Scotch Blackface ewes as Yorkshire Crosses.

[1] In many parts of England half-breds are erroneously called Border Leicesters. In parts of Yorkshire the name Bamshire (from Bamburgh, Northumberland) in used.

Masham ewes are often mated back to Wensleydales in the lower country, producing "twice-crossed" lambs. Both sorts are very useful commercial sheep for winter feeding. The Wensleydale is also the favourite breed for crossing with Herdwick ewes.

The Lincoln (Plate 12, *A)* : Before the era of live-stock improvement for meat production there existed in Lincolnshire a large, coarse breed of sheep, with a remarkably good fleece of long, strong, lustrous combing wool. For long the breeders of the Old Lincoln sheep held out against the introduction of Dishley Leicester blood, but eventually crosses were made, and the modern Lincoln undoubtedly owes something to the outside blood, which, although it temporarily diminished the size of the sheep and the length of wool, improved the fattening qualities and symmetry.

The Lincoln is the largest British breed. Mature rams in show condition occasionally reach 400 Lb. live weight, and the average weight of the wethers (at twenty-two months old) shown at Smithfield in the thirties was practically 3 cwt. The Smithfield weights show the breed to bc about 12 per cent, heavier than the Leicester. Ordinary commercial tegs reach carcass weights of 80 to 90 Lb. at twelve months old. Compared with the English Leicester, the Lincoln is more massively built and has a rather longer and stronger head, which is generally carried a little higher. The ears are always more or less dotted with black. The wool is of great length, that of hoggs frequently exceeding 18 in; it is very dense on the pelt, wavy, lustrous, and strong, but distinctly coarser (about 44/36) than tot of the Leicester. The locks are as broad as two fingers and are less completely separated than in the Leicester. The clip is heavy; 12 Lb. is not an uncommon average in good flocks, while ewe hoggs often give a stone and heavily fed ram hoggs as much as 2 st. of wool. In the show type a good deal of wool is carried on the hind legs below the hocks, and there is a top-knot of long locks.

The Lincoln is quite hardy, commercial flocks being kept on the fairly high and exposed wold and "heath" land of its native county. It is not remarkable for fecundity, 130 per cent, of lambs being considered a satisfactory crop. The quality of the mutton, at least of the older sheep, is generally poor. In Britain, the breed has not spread far beyond its native county, most of the leading flocks being found in Lincolnshire and southeast Yorks, though there are numbers of commercial flocks in Nottingham and elsewhere in the east Midlands. Abroad, however, especially in Australia and Argentina, the breed is in great favour on account of its value for crossing with the Merino. The cross-bred is large and readily fattened, and yields a heavy fleece of useful medium quality wool. The well-known Corriedale breed, originally formed in New Zealand but now widely distributed throughout the world, was produced by fixing the characters of this cross, with some infusion of Leicester blood. In the past fifty years British breeders of Lincolns depended largely upon the export trade, and since this was not very active in the years preceding 1939, the breed was declining in numbers. Under the war-time system of paying flat-rate prices per pound for all fat sheep, irrespective of size, the breed rapidly returned to favour.

The Cotswold, formerly kept in large numbers on the limestone farms of Oxfordshire and Gloucestershire, is still of some importance in the United States, but only two or three British flocks survive. It is nearly as large as the Lincoln and of somewhat similar type but with a somewhat coarser type of fleece.

The South Devon (Plate 12, *B)* is a large, robust type of sheep found in Cornwall and South Devon. In many respects it resembles the Lincoln but is longer in body and, on the average, less squarely and massively built. The wool is dense, long, and curly, but coarser (about 36/32) than in the Lincoln, particularly at the breech. The mutton is of relatively good quality, with a high proportion of lean meat.

The Devon Longwool is smaller than the last (roughly equal to the Leicester in size) but somewhat irregular both as to size and type. The fleece is similar to that of the South Devon. The quality of mutton is again above the ordinary Longwool standard. The Devon is the predominant Longwool breed in North and East Devon and Somerset, but is tending to give way, like other Longwools, to smaller, leaner, and hardier breeds. For fat-Jamb production Devon and South Devon ewes are frequently crossed with Down rams, the Dorset Down, Hampshire, and Suffolk being commonly used.

The Kent or Romney Marsh (Plate 13, *A)* is an old breed native to the rich but cold, exposed, and damp grassland district from which it is named. Like all Longwools the modem breed is indebted to some extent to the Leicester, but it has retained many of its original characteristics and is in many ways the most distinct of the Longwool breeds. Of these it is the smallest, the mean weight being perhaps 5 per cent, less than that of the Leicester. It is low of leg, stout-limbed, and long-bodied. Often it is low at the shoulder and deficient in heart girth in proportion to its development of hind quarter. The forehead is very wide and generally carries some short wool; the nostrils are black The wool is denser than that of any other Longwool and is much finer in quality (50/46) though shorter and less lustrous, and sometimes rather coarse at the breech. The wool is classed as demi-lustre. The fleece is heavy, ordinary commercial flocks yielding an average of 8 or 9 lb., and pedigree flocks usually reading over 10 Lb. The Romney is hardy, does well on grass without hand-feeding, and is relatively immune from such diseases as foot-rot and liver-fluke tot so frequently affect sheep on wet land. The sheep graze singly, rarely flocking of their own will. The breed is not notably prolific, giving perhaps 120 per cent, of lambs under ordinary conditions.

The Kent sheep is very numerously represented in the south-eastern counties, although a considerable proportion of the ewes are crossed with Southdown, Hampshire, or other Down rams. The lambs so produced fatten readily and are of excellent quality. Abroad the breed is very popular, notably in New Zealand, but also in South America, Australia, the Falkland Islands, and elsewhere. New Zealand now competes with England in supplying pedigree sheep of this breed to other countries.

The Southdown (Plate 13, *B)* was the earliest improved of the Shortwool type, and was used in the improvement or formation of all the modern Downs, just as was the Dishley Leicester in the production of the modern Longwool breeds. The Southdown has existed for several centuries on the chalk hills of Sussex, and the old type, before the improvers had set to work, was described as a small, dark-faced, and occasionally horned breed, long in the neck, light in the fore quarter, long and narrow in the back, with a light, thin fleece. The hind quarter was, however, well-developed, and the leg thickly fleshed. The first of the great improvers was John Ellman of Glynde, near Lewes, who started farming about 1780 and bred Southdowns with great success for fifty years. He avoided close inbreeding and, unlike Bakewell, he devoted close attention to the fleece and to the breeding qualities. Also, he increased rather than diminished the size of his breed. Ellman's chief successor in the work of improvement was Jonas Webb, of Babraham, Cambridge, who began about 1821.

The Southdown is distinctly the smallest of the Down breeds and attains, at maturity, less than two-thirds of the weight of the Lincoln. Full-grown rams of the better sort, in fat condition, weigh about 200 lb., and good ewes in breeding condition about 130 lb. The general build is very neat, compact, and symmetrical. The head is very wide, short from the eyes down, of an even mouse colour and is wooled over the forehead, but not round the eyes or over the bridge of the nose. The ears are short and are covered with short wool. The leg of mutton is remarkably thick and heavy. The wool is short (about 2 1/2 in.), dense, and very fine (60/56), commanding the top price per pound for British wools. The clip in good ewe flocks may reach 4½ lb., or occasionally more. For quality of meat the breed is unsurpassed, and as a winner of carcass competitions it occupies a place comparable to that of the Aberdeen-Angus among cattle. The lambs mature and fatten very early and produce small fleshy joints with very little bone.

The breed is reasonably hardy, but below the average of the Down type in prolificness. Up till 1939 Southdown rams were largely used for crossing, and the cross-bred lambs, though small, ordinarily commanded a premium over all other market types. The loss of this premium under war-time conditions resulted in a great decline in the demand by commercial sheep breeders for Southdown rams. There had been some revival at the time of writing , but pedigree flocks in Britain were very few. The breed is still in widespread use in New Zealand for crossing with Romney Marsh and Corriedale ewes for the production of choice quality lamb.

The Shropshire (Plate 14, *A)* originated before 1850 in Shrop-shire and Staffordshire. It was produced from local sheep by crossing with the Southdown, and possibly, in earlier times and to a less extent, with the improved Leicester. Of these local types the two most important were those of Morfe Common, near Bridgnorth, and of Cannock Chase in Staffordshire. The new breed first attracted at-tention at the Royal Show in 1853, and in 1859 it was definitely recognized by being given separate classes. Although the type has changed considerably since those times, no further outside blood has been introduced.

In size the Shropshire stands about half-way between the Southdown, which is the smallest, and the Oxford, which is the largest of the Down breeds. The head is wide above, and short, The former breed standard called for wool down to the nose, but a " clean" face, with the wool restricted to the poll, is now laid down. The change is to be commended, since wool on the face tends to collect grass seeds, burrs, etc., thus making for extra work on the part of the shepherd. The face and legs are of a soft black colour, sometimes a little mixed with grey. The ears are short and pointed. The general form is square and blocky, on short strong legs that are wooled almost down to the feet. The fleece is dense, of good length (fully 3 in.), and of typical Down quality (56/50). In good flocks the average clip will reach 7 lb. The wool is relatively free from dark fibres, though some are generally to be found about the head and legs. The Shropshire is very early maturing, and hence, excellent for fat lamb production. The quality of mutton is good, but the carcass lacks something of the smoothness of that of the Southdown, and can easily be made too fat. The fertility is high, 150 to 160 per cent. of lambs being common. The breed is fairly hardy, adapts itself well to a great variety of conditions, and the rams are very useful for crossing with the leaner and later-maturing types of ewe. In Britain it is still most numerous in Shropshire and the adjoining counties. From about 1920 there was a marked decline in numbers; on the one hand the breed lost ground to the Suffolk and Oxford for crossing with long-wool and half-bred ewes and, on the other, as a foundation for commercial ewe flocks, to the Clun and Kerry Hill. Recently, there has been some revival.

The Suffolk (Plate 14, *B)* was formed in the first half of the nineteenth century by crossing the old black-faced Norfolk Horned breed with the Southdown. The former had existed in East Anglia from early times, and was a very active, hardy, and prolific breed producing a light fleece and a very lean carcass. One small flock still survives. By 1850 the new type had been fairly well-fixed, and by 1859 it was recognised locally as a definite breed. In 1886, separate classes were provided for it, for the first time, at the Royal Show.

Though slightly inferior in weight to the Oxford, the Suffolk is one of the larger of the Down breeds. The head is rather long, generally quite free from wool, and, with the legs, of an intense glossy black colour. The neck is rather long, the bone stout, and the legs free from wool. The fleece is of fair length and density and of typical Down quality (58/52). A good average clip would be 7 lb., but some choice flocks reach 9 lb. Dark fibres in the wool are rather common, and the lambs are dark or parti-coloured when born. The mutton is full of lean meat and of very high quality. In carcass competitions in Britain the Suffolk competes strongly with the Southdown and the Mountain breeds. The prolificness is relatively good. On the light-land farms of East Anglia, where food conditions at mating time are often unsatisfactory, a fall of 130 per cent, is perhaps average. When the ewes have access to abundant green food at mating time, 150 per cent, or more is not rare.

In Britain, large numbers of rams are used for crossing with Longwool ewes, the crosses being so good that they form a favourite type for exhibition at fat-stock shows. In the north, they are now perhaps the most popular breed for crossing with half-breds, and are sometimes used on Cheviot draft ewes. The centre of the breed is still East Anglia, but there are many flocks in other districts of England and in Scotland. Canada, U.S.A., South and East Africa, and Chile have been the principal foreign markets.

The Hampshire Down (Plate 15, *A)* owes its origin to the crossing of Southdowns with two old local types, the dark-faced, horned, Berkshire Knot and the Wiltshire, a white-faced, horned, and very light-woolled sheep that is still in existence. The breed was formed in the first half of the nineteenth century, and was first recognized by the Royal Agricultural Society in 1861.

The early Hampshire was nearly as large as the Suffolk, though shorter legged and more blocky. The recent trend has been towards a smaller type more nearly resembling the Southdown. The ears are long and are carried rather low. The wool on the head reaches below the level of and generally surrounds the eye. The shoulder top and breast are generally very wide and the leg is well developed. The wool reaches about 3 in. in length, is of nearly as good quality as Southdown and is very uniform (58/56), but only moderately dense, and dark fibres are rather frequent. A good average clip would be 6 Ib. The face and legs are of a deep, rich, brownish black.

Hampshires are very rapid growers, lambs reaching great weights at weaning. They do not fatten quite so readily as Shropshires, nor do they produce so fine a quality of carcass as the Southdown. The breed is well-adapted for folding on arable land, and is the favourite on the mixed Down and arable holdings on the chalk formation, where the ewes generally lamb in January, and many of the lambs are sold fat in spring. The ewes breed earlier in. the year than those of other breeds, excepting the Dorset Horn and Dorset Down. The Hampshire is not very prolific, a lambing percentage of 120 being regarded as good. In many flocks a large fall of lambs is not specially desired, as twins cannot easily be got ready for the early lamb market. Moreover, it is difficult on the chalk farm to produce fresh, green food in July, and the ewes often go to the ram in poor condition. The heavy head and wide shoulder of the old type discouraged the use of the Hampshire for crossing with the smaller sort of ewes, but recent years have seen a considerable refinement in these points. Abroad, the Hampshire has met with a considerable measure of success in many sheep countries, and especially in North America, where it is now the most numerous of British breeds.

The Dorset or West Country Down was formed by mating Southdown rams with Hampshire ewes, and by crossing the ptoduct with the native Western Down type. It may be described as generally similar to the Hampshire; it is, however, slightly smaller and rather more blocky in build also, it shows more refinement in the head and bone, and has a distinctly lighter colour of face and legs. The wool resembles Southdownin quality (58/54). The ewes breed very early in the year, and of ten, lamb down in November and December.

The Oxford Down (Plate 15, *B)* is another breed that owes its origin in part to the same early types as produced the Hampshire, having been formed between 1830 and 1850 by crossing these with the Cotswold. Some Southdown blood was also introduced. Crossing of Downs and Longwools is often done in commercial flocks, but the Oxford is the only case in which the cross-bred type has been successfully fixed. The breed was first granted separate classification at the Royal Show in 1862.

The Oxford is the largest and heaviest of the Downs. Insome respects it most nearly resembles the Shropshire, but apart from the difference in size the Oxford has a much longer and less dense fleece, and less wool on the head and legs. There is a top-knot of long wool, and the face colour is uniform dark greyish brown-lighter than in the Shropshire. The head, too, is longer, and the ears, longer and thinner. The clip is heavy, reaching an aver-age of 9 Ib. in the better flocks. The wool is coarser than that of the other Downs (52/48), and indeed only slightly finer than Romney Marsh. The Longwool blood shows up, too, in the carcass. As lambs or young tegs, Oxfords give meat of good quality, but at a greater age, say towards a year, the carcass tends to become fat and coarse, and the size of the sheep passes beyond the desirable limit. Oxford rams are very suitable for mating with the older ages of half-bred ewes. The Oxford cross was for long the most numerously represented on arable farms in the north, but in the years preceding 1939, the preference for leaner joints favoured the Suffolk at its expense. The breed is widespread throughout Britain, with its centre in Oxfordshire and Gloucestershire. Abroad it has proved successful in most districts where food is plentiful. The ewes are very prolific.

The Dorset Horn (Plate 16, *A)* is an ancient breed native to Dorset and the adjoining parts of Somerset, Wiltshire, and Hampshire, including the Isle of Wight. The breed is of medium size with white face and legs, and is horned in both sexes. The nose is pink or flesh-coloured, never black as in the Longwools. It is about similar in weight to the Shropshire, but is longer in body and generally more loosely built, with a lighter fore quarter and a somewhat narrower back. The wool is as fine as that of any of the Downs, excepting only the Southdown and Dorset Down, but is rather less uniform in quality (56/50). It is clear white in colour when scoured, and is free from dark fibres, whereas that of most Downs is creamy white, and often contains black or brown: 5 Ib. is reckoned a good aver-age clip for ewes. Lambs are normally shorn in their first summer and clip 2 *Vz* to 3 Ib. The quality of the mutton is good, but the carcass usually lacks the thickness of flesh of that of the typical Down. The breed is quite hardy, and well-suited to the chalk farms of its native district, consisting of rather poor arable and dry semi-upland pastures. The ewes are very prolific, frequently giving 150 per cent, of lambs, and are excellent milkers. Their most striking peculiarity is that they will often take the ram as early as April or May, thus, lambing in October-November. Milk-fed fat lambs may, thus, be produced by Christmas. Dorset and Hampshire Downs have something of the same peculiar habit, but not in the same degree. Occasional sheep of other breeds also breed out of season. Dorset Horn ewes were at one time widely used for the production of "out-of-season" lamb, but the extensive importation of frozen lamb, chiefly from New Zealand (Canterbury lamb), made the practice unprofitable. A proportion of Dorset ewes will breed twice a year, but no satisfactory

PLATE 16

A. DORSET HORN SHEARLING RAM

B.RVELAND SHEARLING RAM

A. SCOTCH BLACKFACE SHEARLING RAM

B. SWALEDALE RAM

PLATE 18

A. SHEARLING CHEVIOT RAM

B. WELSH MOUNTAIN RAM

A. HERDWICK RAM

B. EXMOOR SWALEDALE EWES

or permanent system of this kind has ever been found possible. The ewes cross well with Down rams, such as Shropshire, Dorset Down, and Hampshire, giving excellent lambs for early fattening. The breed has been extensively exported to Australasia and North America.

The Wiltshire Horn (or Western) Sheep is an old breed that at one time was nearly extinct but has been revived. A considerable number of small flocks are to be found in Northamptonshire and Anglesey. The Western is a sheep of good size, somewhat resembling the Dorset Horn, but of rougher conformation. The skin is often spotted with black, and the breed is peculiar in that it produces practically no wool. Such fleece as it possesses generally falls off in spring. The breed is hardy, and thrives well on grass through-out the year. The rams are used, especially in north Wales and the English Midlands, for crossing, the crossbred lambs being quick growers and coming early to market.

The Ryeland (Plate 16, *B)* is descended from a very small and very fine-woolled breed that was common in the west Midlands during the eighteenth century, and probably much earlier. In most districts this type seems to have been bred out of existence by repeated crossing with Leicesters, but in the poor and sandy tract in south Hereford (Ryelands) many of the original qualities were retained, along with something of the size and fattening qualities of the Dishley sheep. The Ryeland is still a comparatively small breed, lighter than the Shropshire but heavier than the Southdown. The face is of a dull white colour with dark nose, and is woolled down to the eyes. The general symmetry is good and the sheep is well-covered with firm flesh. The wool is of good quality (56/50), not so fine as South-down, but deeper in the staple and of a clear white colour. Good flocks give an average clip of about 6 Ib. The carcass is blocky and thickly fleshed, but with rather more fat than that of the Downs. The rate_of growth of the lambs is only fair. The Ryeland is moderately prolific, and thrives on cold, damp, and inferior land with a minimum of hand-feeding. It is kept entirely as a grassland breed, *i.e.* it is rarely folded on arable crops. The home of the breed is still the southern part of the Welsh Border area, but pure-bred flocks are maintained in many parts of England. At one time, the Ryeland seemed to stand in danger of being ousted by the Shropshire,which is its neighbour on the north side; now its chief competitors are the Clun Forest and the Kerry Hill.

The Kerry Hill (Wales) occupies a position intermediate between that of the Down and that of the true mountain breeds. Geographically, and to some extent as regards its descent, it is midway between the Shropshire and the Welsh Mountain. Its home is the Kerry Hills district of Montgomeryshire. In size it approaches the Shropshire. The face and legs are black-and-white, with the black concentrated at the nose and feet, and there is a tuft of wool on the forehead. The wool is of'Shropshire type, but generally somewhat coarse on the breech (56/48). The ewes are very prolific and good nurses. The quality of the mutton is good. The Kerry ram is the most commonly used for crossing with Welsh Mountain ewes, and again the Kerry ewe is often crossed with the Shropshire or Hampshire for the production of fat lambs. Some exportations have been made, but the breed has not yet become generally known abroad. The Breed Society was The Breed Society dates only from 1899, although the breed has existed in very much its present form for about eighty years.

The Devon Closewool is native to the area lying between Barnstaple, South Molton, and Blackmore Gate in North Devon. It is derived from crosses between the Devon Longwool and the Exmoor Horn. Some of the flocks date from the end of the nineteenth century, but it was not until 1923 that a Breed Society was formed. Intermediate in size between the two parent breeds, and hornless, the Closewool otherwise approaches more nearly to the Exmoor Horn, being compact in build, thickly fleshed, and carrying a dense fleece of useful medium-quality wool (50/46). It may be described as like a small Romney Marsh, but it is shorter-bodied in relation to its height. The Closewool is a hardy breed, admirably suited to semi-upland conditions and living on grass for the greater part of the year.

The Clun Forest is an old local breed which originated in the Clun district of Shropshire, and was formerly restricted to the hilly districts of Shropshire, Radnor, and Montgomery. It is related to the Kerry Hill but carries more Shropshire blood. The breed is dark-faced, hornless, of medium size and, like the Kerry Hill, is notably hardy and prolific. It may be said to be intermediate between the Down and mountain types. The weight of fleece is about 6 lb. and the quality of the wool is similar to that of the Ryeland.

In recent years the breed has risen in popularity very rapidly, and it is now to be seen in large numbers not only in the Welsh Border country but in the Midlands and south. The ewe flocks thrive on pastures with little hand-feeding, and are said to require very little shepherding. The wether tegs are frequently fattened on roots and settle down well in folds. If the ewes are mated to Down rams-e.g. Hampshire or Suffolk-the lambs fatten readily on leas and reach good weights.

It is common practice to put Clun ewe lambs to the ram at an, age of eight or nine months. The usual expectation is that about 60 per cent, will conceive.

The Radnor is native to its name county, and in its economic characters resembles its geographical neighbours, the Clun and the Kerry Hill. The original breed was tan-faced like the Welsh Mountain, rather small, and very hardy. Some crossing with the Kerry Hill and Shropshire increased the size with some sacrifice of hardiness. There was, at the time of writing, no breed society.

Mountain Breeds : The Blackfaced mountain type seems to have been first known in the Pennine Chain, and all the modern varieties of it appear to have a common origin in this district.

Of these the **Scottish Blackface** is now the best known and is the most numerous of all British breeds. It must have reached the southern uplands of Scotland several centuries ago, but was not introduced into the Highlands till about 1770. Previous to this time the Highland grazings were stocked mainly with cattle, and what few sheep there were belonged to the old Celtic soft-woolled type, which is still to be met with on some of the Western Isles. The Scottish Blackface (Plate 17, *A),* when reared in a true mountain environment, is a small sheep, but when heavily fed for fat-stock shows it reaches a weight greater than that of the Southdown or Ryeland. The face and legs are either black or "brockit," *i.e.* black with white patches, the latter being preferred, though excess of white is objectionable. The head is rather short and, in show specimens, very deep and strong at the nose. The horns of the ram are large, coming out level with the head and forming a wide spiral. The fashionable type of wool is long, nearly reaching the ground, and of a coarse and hairy character (40/28) but free from kemp. Some stocks have a finer and softer wool that can be used for rough over coatings and blankets but most of the wool goes for carpet manufacture. The average-lip under mountain conditions is probably about 4 lb., and the best flocks rarely exceed 5 lb. The general conformation has been immensely improved in recent times, and the modern sheep has a good back and leg of mutton, though the wither, as in all mountain types, is often narrow and sharp. The quality of carcass, whether as lamb or mutton, is first class. The tail is naturally shorter than in other breeds, reaching only to the hocks; as in other mountain breeds, it is frequently left uncut.

The Blackface occupies practically all the heather-clad or " black" hills of Scotland and Northumberland upto altitudes of 3000 ft. The highest grazings formerly carried wether stocks, the male sheep being kept on till they were two, three, or even four years old, but many such stocks have been displaced either by breeding ewes or by deer. Most of the lambs not required for breeding are now sold either fat from the hills in the latter part of summer, or as stores in autumn. On the higher grazings the Blackface is kept pure, while on the lower hills the ewes are crossed with Border Leicester or, in some districts, with Wensleydale rams. The cross lambs, known in the first case as "Greyfaces", "Mules," or "Crossbreds", and in the second as "Yorkshire Crosses", make excellent mutton at from seven to fifteen months old, though some are sold as fat lambs. The Blackface is exceedingly hardy. Generally, the ewe hoggs destined for breeding are sent to the low country to be wintered (on grass only), but the ewes live on the mountain throughout the year and get no artificial feeding except a little hay, and that only when the ground is covered with hard-frozen snow. The ewes are prolific. Under hill conditions twins are not generally wanted, but good draft ewes when taken to the low country give upto 140 per cent, of lambs.

The Lonk is found in the mountainous district of west and south-west Yorks, east Lanes, and north-west Derbyshire, It is a larger, longer-bodied, and rather longer-legged sheep than the Scottish Blackface, with often more white in the face. The tail is long. The fleece is considerably finer (50/44) and both denser and heavier (6 to 7 lb.) than that of the Scottish. The quality of the mutton is excellent, but the breed is distinctly less hardy than either the Scottish Blackface or the Swaledale. Wensleydale and Leicester rams are those principally used for crossing, where crossing is done.

The Swaledale (Plate 17, *B),* another useful and hardy breed, is also somewhat larger and longer-bodied than the Scottish Blackface. The fleece is shorter and finer in front but is coarse on the breech, and kemp is quite as common. The face is black or dark grey with a "mealy" (grey) nose, and the legs are spotted. There is usually a good deal of black wool on the nape and throat. Swaledale lambs are said to fatten rather more readily than Scottish Blackfaces and the ewes to milk rather better, but the hardiness is probably not so good. This breed is found from Swaledle westwards to the Pennines and into Westmorland. Where crossing is done the Wensleydale ram is used, and the female progeny, known as Masham ewes, are a common type of breeding stock on grassland farms.

The Rough Fell, also known as the Kendal Rough Sheep, is a closely related breed found in an area centred on Kendal and embracing parts of the West Riding of Yorkshire and of Westmorland. It is larger, somewhat rougher in conformation, and shorter woolled than the Scottish Blackface, and shows a brownish tinge in the face. It is regarded as a hardier sheep than either the Swaledale or the Lonk.

The Dales-Bred is a rather larger type with a heavier fleece. It is found on the lower and better grazings in the same area. The wool approaches that of the Scottish Blackface in quality (40/32) and the ewe fleece averages about 5 lb. Older ages of ewes are com-monly crossed with Wensleydale or Teeswater rams.

The Derbyshire Gritstone, native to the high Peak district of Derbyshire, has the softest wool (50/46) and the closest fleece of all the breeds of the Blackface type, and is further distinguished by the absence of horns. The face and legs are mottled with black and white.

The Cheviot (Plate 18, *A)* is regarded as derived from the an-cient Celtic tan-faced breed, and would thus be related to the Welsh Mountain and to the old soft-woolled sheep of the western Highlands and Hebrides. It has existed on the hills of southern Scotland from very early times, though formerly the breed was much smaller and probably finer-woolled than now. There are some records of crossing with Lincolns and Leicesters about the end of the eighteenth century, but the improvement in size and in mutton qualities has been accomplished mainly by selection. The Cheviot is of medium size, reaching, when intensively fed, about the same weight as the Shropshire. The size of hill-bred sheep naturally depends on the abundance and quality of the herbage. Mature rams of the best class, in show condition, weigh over 200 Lb. The head and legs are white, with occasionally a shade of tan, and covered with rather hard, short hair. The head is rather short and very strong through the nose; the ears short and carried very high, giving the animal a very characteristic alert appearance. Short horns still occasionally appear-perhaps to the extent of 5 per cent.- in males. The tan colour of face is gradually being eliminated. The wool is of fine quality (56/48) but with a harsher handle than typical Down wool, of medium length (3 to 4 in.), and quite dense on the pelt. It is largely used for the manufacture of tweeds. Kemp is not common, but does occur. Mountain flocks clip from 3 '/2 to 5 Lb., and the better sort of ram hoggs reach 10 or 12 Lb. The general conformation is good. The body is distinctly long and the back sometimes slack; the hind quarter is generally good, but the fore quarter, as in most mountain breeds, tends to be proportionately light and the shoulder rather sharp. The mutton quality is excellent, and the breed and its crosses (especially those from Southdown and Suffolk rams) are often very successful in carcass competitions. The Cheviot is hardy, but distinctly less so than the Blackface. In Scotland and the north of England, it is chiefly to be found on the lower and more grassy hills, the heather land being given over to the Blackface. Cheviots are also kept as "park" sheep, *i.e.* on semi-upland enclosed land, and are then wintered with some turnips and other artificial food. The breed is prolific under good conditions and the ewes are excellent milkers. Hill Cheviots are very active and difficult to confine.

There are two main centres of the breed, the one in southern Scotland and the north of England and the other in Caithness and Sutherland. North-country sheep are larger and leggier, somewhat less compact in build, and longer of head and ear. The ears, also, are carried lower. They have finer but rather more open coats and tend to be less well clad over the belly. They require rather better food conditions than the true hill type of the Border area. A separate flock book for North Country Cheviots was established in 1946.

On the better hill farms the older Cheviot ewes are crossed with Border Leicesters, and on certain farms only half-bred lambs are produced, the Cheviot ewe stock being replenished from outside sources. Half-bred ewe lambs are in very keen demand for breeding, and command a much higher price than wethers of the same cross. Half-breds produced from north-country Cheviot ewes are larger and more prolific than the corresponding progeny of Border ewes, though they require somewhat better food conditions. Draft Cheviot ewes are mated either with Border Leicesters or Suffolks. Cheviot ewes and ewe lambs are often taken far south into England for breeding purposes, and there are colonies in Brecon and on Exmoor. The breed has met with some success in Canada, the United States, and New Zealand.

The Welsh Mountain (Plate 18, *B)* is an original Celtic breed that has been kept almost free from extraneous blood. It is the smallest of the breeds here described, reaching about half the weight of the Leicester or Oxford and not much more than two-thirds that of the Southdown. The Welsh sheep is rather long-bodied, light in the fore quarter, and the face and legs are either all white or have patches of a light tan shade-the latter being commoner and preferred. The rams only are horned. The breed embraces several types, including a larger improved type on the more fertile farms with some low ground and smaller types on true mountain land. The wool of improved sheep is short, fine, and dense, comparable to Shropshire or fine Cheviot in quality (54/46), but that of the true mountain type is often coarse at the breech and contains much kemp. An average clip for ewes is 2 '/2 lb., rams reaching 5 or 6 lb. The mutton is comparable in quality to that of the Southdown, Cheviot, or Blackface, and indeed is regarded by many judges as the finest procurable.

The Welsh is a hardy breed, living on the higher mountains during summer and being brought down to lower levels for wintering and lambing. The sheep are very active and rather wild, and hence are very difficult to confine. The ewes are excellent breeders and nurses. Draft ewes may be mated with Southdown rams to produce choice small fat lambs. For larger fat lambs the Wiltshire Horn is preferred. Where the produce is to be carried on to the teg stage, Kerry Hill or Clun rams give good results. The breed is found throughout the whole of the mountainous district of Wales, the population being densest in the north. A black strain of the breed exists, but is generally kept in lowland parks.

The South Wales Mountain is a rather leggy type with a very kempy fleece, suitable only for carpet manufacture. It is found in Glamorgan and Monmouth and in parts of Carmarthen and Brecon. It is the largest of the Welsh types.

The Herdwick (Plate 19, *A)* is a breed of rather striking peculiarities found in the bleak, exposed tells of the Lake District. Its origin is quite unknown, although there is a traditional that the foundation stock came from a Scottish or Norwegian vessel that was wrecked on the Cumberland coast two or three centuries ago.

The size is rather less than that of the Blackface mountain type, though varying a good deal with the quality of grazing. The rams are generally but not always horned, while the ewes are hornless. The face and legs of the lamb are of a deep blue-black colour, becoming grey and sometimes eventually white with age. The wool, particularly that on the belly, is often mixed with grey; it is long and coarse (40/32) and contains much more kemp than that of the Scottish Blackface. Ewes clip from 3 to 5 Lb. according to conditions, and rams upto 9 or 10 Lb. The shoulder top is generally sharp and the breed is late maturing, but the mutton is of fine texture and flavour. The ewes are sometimes not made to bear lambs till they are three years old-a year later than the normal age. The Herdwick is generally regarded as the hardiest of all British breeds; the sheep subsist largely on heather and juniper, and survive long periods of storm. Draft ewes taken to the low country are generally crossed with Wensleydale, Leicester, or Border Leicester rams, and the progeny make excellent butchers' sheep. The Wensleydale is the ordinary cross under hill conditions. The breed is declining in numbers owing to the progress of afforestation in its home area.

The Exmoor Horn or Porlock (Plate 19, *B)* is an ancient breed native to Exmoor and the Brendon Hills in North Devon and West Somerset. It is of small size, horned in both sexes, with white face and legs and a black muzzle. The fleece is of the Cheviot type, but rather longer and stronger and coarser at the breech (56/40). Ewe flocks give clips upto 5 Lb. The general conformation is compact and neat, with a peculiarly rounded symmetry. There is rather frequently a deficiency behind the shoulder. The breed is hardy, but the ewes generally get some roots and hay in winter. The mutton is of good quality.

The Dartmoor, like the Oxford Down, Kerry Hill, and Clun, cannot rightly be classified in any of the three main groups of breeds. Though originally a Moorland breed, the Dartmoor has been so increased in size and improved in form that it must now be considered a Longwool. Like the true Longwools it owes something to Dishley Leicester blood.

It is larger than any of the mountain breeds proper, and hornless-short stubs being seen exceptionally in rams. The face and ears are white with black spots, the latter generally somewhat concentrated towards the nose. The fleece is long, curly, and lustrous, of coarse Longwool quality (36/32), and the clip is heavy. The sheep is long-bodied, strong-limbed, and well-covered with firm flesh. The ewes are wintered on grass and hay, and lamb in the beginning of March. The breed is rather late-maturing but produces yearling wethers of fine quality. Older ewes are crossed either with Down or with South Devon rams. A white-faced strain is separately registered.

The Shetland is an interesting breed, so far little improved, but with certain potentially valuable characteristics. It is very small, slow-maturing, and thin-fleshed, though the texture and flavour of the mutton in well-fed sheep are good. The colour is black, grey, or brown (moorit), often with white patches, or entirely white. The white and moorit wools command the highest prices. The fleece is light and of very mixed quality but contains a proportion of very fine soft wool which commands a higher price than that of the Southdown. In Shetland the sheep are not shorn, the loose wool being picked off at intervals and the finest sorted out for the local industry of shawl-making. The breed, like others in Northern Europe, is naturally short-tailed. The rams are horned, the ewes generally hornless. At one time the breed occupied the whole of the Shetland Islands, but it has given way to some extent before the Blackface. Crosses with the Cheviot are useful both for wool and mutton.

COMMERCIAL TYPES AND CROSSES

It will be clear from references in the preceding pages that certain areas of the country breed far more sheep than can be fattened locally, and that some districts will have surpluses of ewe lambs suitable for breeding. The surplus is sold, and is moved into areas, where the flocks are not self-supporting. The following are the more important classes of stores and breeding animals that pass through markets :

Store Lambs : A variable but rather small proportion of mountain wether lambs are sold to butchers direct from the grazings where they have been bred, but the bulk are marketed as stores in August and September. The main sources are the Highlands and southern uplands of Scotland, the northern counties of England, the Pennine region, and Wales. The mountain breeds have a limited capacity for growth, and in any case are not very suited for folding on roots, hence, the great majority are fattened on pastures, aftermaths and rape, and reach the fat market mainly from October to December. They yield small carcasses of between 25 and 35 lb., which in normal times command a high price per pound. Ewe lambs are in demand for breeding, and in most areas only culls can be bought at prices which make it possible to fatten them at a profit. The typical product of the lower hill areas in the same regions and in Caithness and Sutherland is a cross between a mountain ewe (Blackface, Cheviot, a Longwool ram. The Border Leicester x Cheviot ewe lambs (Hybrids) are in great demand for breeding, and few are fattened as tegs; but the wether lambs are a fairly important class of store. The breeders' demand for Longwool x Blackface ewe lambs (Greyface, Mule, Masham) is not so great, so that many of these, as well as the wethers, are available for fattening. Greyface and similar crosses are commonly run on lowland pastures for the earlier part of the winter and finally fattened off on roots, hay, and concentrates, reading the fat market mostly from March till May; they produce choice carcasses of the size (50 to 60 lb.) that the consumer prefers at that time of the year.

Apart from mountain and hill areas, the chief sources of store lambs are: (1) Part-arable farms in southern Scotland, Northumberland, and Cumberland (crosses between Suffolk or Oxford rams with Half-bred or Mule ewes). (2) The Welsh border area (Kerry Hill, Clun, etc., and crosses of these breeds with Down rams). (3) Kent (Romney Marsh wether lambs and crosses between the Southdown ram and the Romney Marsh ewe). (4) The chalk area, from Dorset to Norfolk (mainly Downs, such as Hampshire, Suffolk, and Hampshire-Oxford crosses). The lambs from these four areas are naturally available earlier than those bred on the hills, the largest sales taking place in June and July. The more forward lots are fattened off on aftermath, rape, etc., but the majority are folded on roots and provide the bulk of the supply of mutton from December till March, yielding carcasses of about 70 to 80 lb.

Ewe Lambs : The mountain-sheep farmer normally retains the best of his ewe lambs for his own flock, selling his "seconds" to farmers in lower Mil areas, who use them to produce cross-bred lambs. As already noted, there is a sale for some of the cross-bred progeny for breeding, the demand for Half-breds being especially strong. Many downland fanners also make up lots of "second" ewe lambs which are bought by other breeders.

Young Ewes : Some breeders hold over their surplus ewe lambs and sell them at about eighteen months of age (theaves or gimmers), when they are ready for breeding. There are also some specialists who have suitable areas of lowland pasture and who buy ewe lambs and sell them as the aves a year later. Many breeders in the Welsh border counties keep all but the poorest of their ewe lambs and sell these females at two and a half or three and a half years old.

Draft Ewes : Under comparatively rigorous conditions for any particular breed the death-rate among ewes tends to rise after about five years. On the other hand, rich land tends to make ewes too fat for breeding. Hence, it is often advantageous to draft ewes from poorer to better farms at "three-crop" (four and a half years old) or "four-crop" (five and a half years old). The latter are often fattened with their next crop of lambs, but the former may produce two further crops of lambs in the second flock. There is a large demand for draft ewes, the sales mostly taking place shortly before the breeding season, *e.g.* in May for Dorset Horns and Dorset Downs, and in September to October for mountain breeds.

SHEEP BREEDING, REARING AND FEEDING

It seems necessary to begin this section with an explanation of the names that are applied to sheep of the different sexes at various ages. Age in sheep is reckoned from the time of shearing, which in this country is normally done for the first time at the age of fourteen to sixteen months, and at yearly intervals thereafter. The term *lamb* is not very precisely limited, being sometimes conpned to the period before weaning and sometimes extended to cover the first six, eight, or occasionally ten months of the animal's life. However, at some time in autumn, lambs become in the north *hoggs* or *hoggets* and in the south *tegs*. The sex is also denoted, males being known as tup or ram lambs, tup hoggs, ram tegs; castrated males as wether or wedder lambs, wedder hoggs, etc.; and females as ewe lambs, etc. in some districts the term *chiver* is used instead of ewe lamb. After shearing we denote the males as shearling rams or tups and shearling wedders, respectively. The terms *diamond tup* and *dinmont* are also applied at this age. The female at shearing becomes a *glimmer* (north) or *theave* (south). Thereafter we have two-shear rams, wedders, or ewes. Ewes may also be distinguished according to the number of crops of lambs they have borne, as two-crop, three-crop, etc., the first lamb being borne at two years old. In some districts, it is customary to denote age by the teeth, as two-tooth, four-tooth, etc., but these terms are less satisfactory on account of the variation in dentition according to breed and environment.[1]

The average period of gestation in sheep is about 147 days, varying a day or two with the breed. The ordinary practice is to complete preparations for lambing about a week before the elapse of five calendar months from the time of turning out the ram. Females are normally mated for the first time as gimmers, to lamb at two years old. Hoggs in good condition will frequently take the ram at an age of seven or eight months, and will then lamb at little over a year; but breeding so early is unprofitable unless the food conditions are really good. As it is rare for more than 70 per cent, of the hoggs to conceive, and as these are apt to suffer in their development, the practice is comparatively rare, and is only to be recommended with well-grown sheep under favourable conditions. The only region where it is common is the Welsh Border. In-lamb tegs should receive specially liberal feeding during the last month of pregnancy. Mountain gimmers, if they are small and ill-nourished, are sometimes given an extra year to develop, lambing for the first time at three. Ram lambs, especially those of the quicker growing breeds like Hampshires, Oxfords, etc., are often used for breeding at seven months old or little more. They should, however, be allotted a smaller number of ewes ton old sheep, generally not more than thirty or forty.

---

[1] In sheep the permanent incisor teeth usually erupt as follows, though heavily fed lowland sheep get their teeth earlier, and mountain sheep later;

First (central) ... ... 12 to 15 months

Second ... ... 18 to 21 months

Thirde ... ... 24 to 7 months

Fourth ... ... 33 months

Shearlings and older rams, so long as they remain active, may have from forty to fifty ewes under mountain conditions, and from sixty to eighty when the flock is in enclosed fields. Where it is desired to make the fullest use of a specially good ram, the ewes, as they come in heat, are taken one by one to the ram for a single service. A "chaser" ram is employed to seek out the ewes that are in season. A piece of sacking is tied round the chaser's middle to prevent him from serving.

The time of mating is limited in part by the time of occurrence of heat in the ewe. Thus, Dorset Horns will frequently take the ram in May, Dorset Downs and Hampshires as early as June or July, while mountain ewes rarely come in heat till late September or even well on in October. In most breeds the onset of cestrus is governed mainly by the hours of daylight-i.e. the ewe begins to come in season with the shortening day. The control of the cycle is by the pituitary body, and it is possible, by injecting pituitary extracts, to make ewes breed at any season. Apart from this the time of mating is fixed by the breeder in accordance with the average climatic and food conditions prevailing in spring. Thus, in the south of England, on arable farms, lambing may begin in December or occasionally earlier, while on Highland grazing it does not commence till late in April. The customary dates for turning out rams in various flocks are usually closely adhered to from year to year, and vary from the end of July (early Hampshires) to mid-September (Lincolns), mid-October (Half-breds), 10th November (average Mountain Cheviots), and 22nd November (Blackface Mountain). Rams are left in the flock for about six weeks, which is ordinarily long enough to ensure that practically all the ewes will conceive. The period of cestrus in the ewe lasts about twenty-seven hours and recurs, if she does not conceive, after about sixteen days. A six-weeks' breeding season thus covers at least two and, in a majority of cases, three cestrous periods. In lowland flocks, where the ewes will have close attention at lambing, it is necessary to record the dates of service. To this end the ram is smeared between his forelegs with colouring matter so that he leaves a mark on each ewe that he serves. At the end of, say, each week, the ewes that have taken the ram are given a distinguishing mark, which remains till spring as an indication of the date of lambing. At the end of sixteen days the colour on the ram is changed to a darker one, so that ewes not holding to the first service can be immediately known by the fact that they are marked with two colours. Any such should be examined, and if necessary, clipped about the tail and inside of the thighs. The longer-woolled breeds are generally so clipped before tupping begins. Occasional ewes that have failed to hold to the first service should be placed with another ram in case the second service should also have proved ineffective. If many ewes are being twice marked another ram should be procured; in large flocks it is best to run two or three rams together with a combined allotment of ewes, as the accident of one ram proving infertile will then cause comparatively little loss. If, however, the rams fight, this plan must be abandoned; also, the method is not applicable in pure-bred flocks, where matings are made with close regard to individual merits and faults. Under ordinarily good conditions and on low ground die proportion of barren or "eild" ewes should not exceed I or at most 2 per cent., in hill flocks it may reach 3 or 4 per cent, in ordinary seasons.

The number of twins, or at least of double conceptions, depends partly on the inherent fecundity of the ewes and partly on their treatment before and during the mating season. The smallest number of lambs will be obtained when the ewes are lean and weak, or very fat, or are going back in condition. The highest birth-rate, on the other hand, is obtained, if the ewes at mating time are in moderately lean but rapidly improving condition This is assured-if the sheep have been in low condition after the previous weaning-by bringing them on slowly till within a week or two of tupping, and then by "flushing" on rape, young seeds, or other fresh nutritious food. "Flushing" causes an increased proportion of the ewes to breed early, *i.e.* at the season when there; is the greatest chance of multiple ovulations. Failing a supply of fresh green food the ewes may be given up to a pound per head of concentrate daily, avoiding specially fattening materials. Oats are very suitable. The proportion of multiple ovulations *i.e.* of heats at which two or more eggs are produced) is highest in the earlier part of the breeding season, this explaining the well known tact that more twins are usually born at the beginning than' at the end of the lambing season. Delay in mating after the time when the ewes would begin to take the ram will, thus, reduce the proportion of twin births. Sometimes twins are not wanted-as, for example, in mountain flocks; ram breeders, too, if they sell their output as lambs, do not want a high proportion, as a good' single ram lamb is often worth more than a pair of moderate twins. Even, in ordinary commercial flocks there is often a limit to the number of twins desired *e.g.* on account of a limited amount of good pasture and in such cases heavy flushing is only resorted to in years when the ewes are obviously backward. Draft ewes, which are usually sold off the hill in autumn, to be kept for a further one or two years under lowland conditions, will generally be wanted in lamb earlier think the normal time, and must be brought forward in condition more rapidly than the rest. To this end their lambs are often weaned earlier than those of the oilier ewes. Rams often require special attention during the breeding lesson. Longwool or other lowland rams put out among mountain ewes must receive a feed of concentrates once a day at least, otherwise they will very rapidly become reduced in condition.

In countries where the winter is very severe, such as the northern United States and Canada, breeding ewes are housed in cool barns during winter, going out for exercise only. In this country they are always wintered out of doors. Ewes of the hardier mountain breeds are left on their usual pastures, being gathered to the lower and sheltered areas only at the prospect of snowstorms. Generally they are made to live without hand-feeding unless, or until they are in rather severe straits, which is only in. the wildest weather, or when the ground is covered with hard-frozen snow. When, 'given hay' unnecessarily they tend to stop foraging for themselves, and since the supply of hay on hill farms is necessarily very limited, they are then in worse plight than if they had been left alone. This is particularly true, if feeding is done during the early part of the winter. Instead of relying on hay to a large extent the hill shepherd endeavours to save the lower areas of the pasture during late summer and autumn, so float he has a reserve of natural herbage for winter food. Weak ewes, however, must be taken in from the hill and kept in enclosures near the homestead, where they can receive whatever special attention or feeding- is necessary. On lowland pasture-farms the sheep may run on grass without any additional food during the early part of the winter. Later on they are given upto 2 or 3 lb. of hay, according to the condition of the pasture, and sometimes, from a month before lambing, a moderate ration of cake and corn.

On arable or semi-arable sheep farms the winter management varies greatly according to breed, the amount of roots, etc., available, and the time of lambing. Half-bred ewes in the north due to lamb in March and

April, are usually am on grass without extra feeding until towards Christmas. As the pasture gets bare they are given a limited ration of turrlips, say 1 ton per hundred ewes per day, or 11 lb. apiece. Later they will generally be folded on turnips for half the day, or overnight, returning to the pastures at other times and receiving perhaps a little hay in addition. When roots are abundant ewes are sometimes folded permanently on the crop; but the method is not always satisfactory, the ewes, though they may keep fat, getting into a soft and unhealthy condition. In any case they should be put back on pasture, with a restricted root ration and some hay, a few weeks before lambing is due to begin; they also benefit by daily exercise, such as may be given by driving the flock daily to and from an outlying field. A light daily ration of concentrates will be given from a month before lambing in order to bring the ewes into good condition for milking. In some districts chaffed oat sheaves, straw and grain together, are fed in lieu of hay and concentrate. In March, just previous to lambing, the ration might be: turnips, 20 lb.; oats, peas, dried grains, 1 lb.; hay, 1 ½ Lb. varying, the hay particularly, according to the amount of rough herbage on the pasture fields. Draft ewes in the lower districts, that are due to lamb perhaps a month earlier, and are to be fattened along with their lambs, are given somewhat more liberal treatment.

In the south on those few farms where folding is still practised, a common succession of winter foods is: (1) turnips, (2) swedes and marrow-stem kale grown in alternate strips, (3) swedes and thousand-headed kale, and (4) swedes. Folding is done for a longer period and more regularly. A few mangolds are often valuable, after lambing, in tiding over the flock until" seeds" or other summer food is available. Clover, lucerne, or sainfoin hay, or pea haulm, may replace the seeds hay of the north. The aim is, however, the same, viz. to keep down costs during the early stages of pregnancy and, by adding concentrates later, to ensure strong lambs and a plentiful flow of milk. In one series of experiments, it was found that halt-bred ewes which gained 20 Lb. of live-weight during their period of pregnancy produced quite as large and as vigorous lambs as others which were more heavily fed and gained 50 Lb. in weight. On the other hand, a third group which were more poorly fed, so that they barely maintained their weight, produced a large proportion of very weak lambs, many of which could not suck.

On sheep farms where the arable land is heavy, as, for example, in many districts at Northumberland, folding in winter is impossible. In such cases the *root* crop is stored in autumn and the daily allawance is thrown out to the ewes on grass.

Before lambing, the ewe flack should be collected and looked over to see that the teats are clear of wool, otherwise the lambs will be liable to seize and swallow locks of wool in their attempts to suck Gimmers (theaves) often have to have a considerable amount of wool removed. The operation is known as "udder-lacking."

Lambing, in hill flocks, goes on in the apen, and generally on the mountains, though in some cases the ewes are brought down to lower enclosed pastures. It generally begins in the second or third week of April and continues throughout the greater part of May. Each shepherd, who will have a "hirsel" of perhaps twenty-five or thirty score of lambing ewes under his charge, is given a temporary assistant (or at least a half share of the time of such an assistant) for the busy and critical period of four or five weeks. This arrangement makes it possible for the sheep to be seen at learn three times a day, and will generally ensure that distance; is available in cases of difficult or protracted labour. The chief causes of heavy martality are snow or heavy rainstorms; in the first case the lambs may be buried, or simply exhausted by the cold and the heavy going; in the latter they are very liable to be drawned in attempting to follow their dams over swollen torrents, in bad weather the sheep are gathered into the naturally sheltered areas on the lower ground. Even in good lambing seasons weak ewes and these with weak lambs require to be brought down and fed, but the main part of the flock should be made to travel the ground in a regular way. The natural habit of hill sheep is to spend the night on the hill-tops, to move down to the valleys in the morning and up again in the afternoon, grazing as they go. When the lambs are young, however and especially if the ewes are in low condition-many will tend to' lag behind and to remain on the lower ground, which would then soon become foul with drappings; very heavy warm infestation may result Hence the shepherd must see that all that are able do actually go up the heights in the evening, and this often entails heavy work. The earliest herbage on mountal grazings is the "Draw Moss" (sheathing cotton-sedge or cotton grass), *Eriophorum vaginatum* which grows on marshy land. On grazings where this is not abundant, care must be taken to preserve it by avoiding aver drainage, and the sheep should have full access to the marshy areas in early spring.

A considerable part of our mountain grazing in Scotland and Northern Ireland is infested with sheep ticks *(Ixodes ricinus)* which carry two specific diseases (Louping-ill and tick-borne fever) and also cause loss by producing abscesses and blood poisaning (pyremia) in young lambs. The ticks are most common an coarse, dense-bottomed herbage where the humidity is high; hence, any treatment that will promote a finer sward and reduce humidity will help to control the pest Drainage, the employment of cattle to eat off the coarse stuff, and burning are all helpful, while repeated dipping of the sheep tends to kill off the parasites. Lambs are most susceptible to tick-borne disease, and it is best to try to protect them. Dusting them with an insect-powder, such as derris or a preparation of D.D.T., is easier than dipping and gives about four weeks protection.

Considerable areas of hill-grazing from Bodmin Moor in Cornwall to the norm of Scotland are deficient in cobalt Cobalt deficiency leads to reduced numbers of lambs, which make very poor growth. The areas in question have been recognized in the past as "pining" land, but the condition has been diagnosed only in recent years. Good results have been obtained by applying cobalt sulphate, at the rate of 2 lb. per acre, to the pastures and also by providing salt licks containing cobalt. The more certain preventive is to drench the ewes in autumn with 200 to 400 mg. of cobalt, and to drench the lambs in spring and again in summer.

Ewes on mixed farms will have been classified in autumn into five graups according to the period (week) in which they are expected to lamb. Each batch in succession, as it comes due to lamb, is removed to the lambing field, which should be a good dry pasture with the best available natural shelter. Artificial wind-breaks, consisting of thatched hurdles or sheep-nets stuffed with straw, and set up in the form of a cross, should be placed here and there. In or near the lambing field should be a lambing shed with a hut in which the shepherd and lambing man spend alternate nights. The shed is usually a permanent structure; some farmers prefer temporary shelters on the ground that they are less liable to harbour disease, but permanent erections, if they receive a thorough annual disinfection, are not open to serious abjection. The shed consists of a series of pens, about 4½ or 5 ft square, with a low roof and arranged to form a hollow square (Fig. 12). In ordinary cases the central yard is open, but for valuable pure-bred flocks the whole may be roofed in. In any case, the shed should be on dry ground with a sheltered situation and a southern exposure.

During the day the ewes go out in the lambing field; at night they are confined in the central yard of the lambing shed, which should be well littered with dry straw. Ewes lambing at night are placed one in each small pen. Assistance in lambing should not be given too soon, and the assistant should thoroughly disinfect his hands before and after each

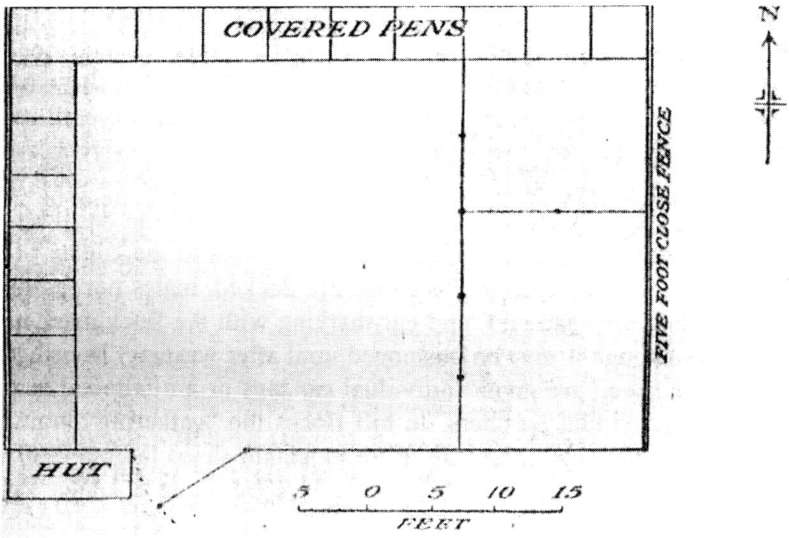

FIG. 12 : Lambing Pen for Flock of 300 Ewes

case. Each ewe and her lamb or lambs maybe given a distinguishing dot of colour, in order to avoid any possibility of subsequent confusion; but as a ewe generally keeps in touch with one of her lambs it is often considered sufficient to mark the twins. Generally there will be a proportion of triple births, as well as a few orphan lambs. On the other hand, there will be many singles and a few ewes whose lambs die. There should be a re-assortment, by transferring lambs, so that the flock is made to consist entirely of singles and doubles. In cases where a lamb has died its skin should be tied over the back of the stranger, when the latter will generally be adopted at once. Other devices, such as that of rubbing the ewe's nose in whisky to deprive her temporarily of her sense of smell, are successful in other cases. Refractory ewes may have to be tied up by the neck for a few days in order to prevent maltreatment of foster lambs. In any case, re-asserted families must be kept penned together until it is seen that they have settled down. Gimmers (theaves) should not be given second lambs, if ewes are available.

The ewes and young lambs are kept in the more sheltered fields near the centre of operations for a week or two, and are then put out in lots to the ordinary pastures. On enclosed farms the flock will be split up into lots of similar age, etc., gimmers with double lambs being given the best grass, ewes with singles the worst. Feeding with roots, hay, and concentrate rust be continued until grass or a forage crop becomes available. This may be as early as mid-April in the south, often after mid-Main the north. When grass becomes plentiful, roots and hay and withdrawn, and the concentrate is either withheld or is transferred from the ewes to the lambs. Soon after lambing is completed the lambs are docked, males not required for breeding are castrated, and ear-marking with the flock mark may be done-although it may be postponed until after weaning. In pedigree flocks the sheep are given individual ear tags or are tattooed in the ear with individual numbers. In hill flocks the "gathering" is taken as an opportunity of enumerating. In mountain sheep the tail is often left long (for protection), but all lambs that are likely at a later stage to receive roots or concentrates are better docked. Males that cannot be castrated "chasers" or "rigs" have the tail left long as a distinguishing mark.

Lambing percentages vary greatly with conditions. Mountain flocks in normal years may give from 60 to 120 lambs per 100 ewes-less on the poorer and higher ground, more on the lower and better sort of grazings. In years of storm and flood the increase may be as low as forty or fifty, and occasionally the ewe lambs will barely be numerous enough to replace losses in the older sheep. Probably the average of lowland flocks all over the country is about 130 per cent., and the most prolific flocks reach an average, over series of years, of about 160.

In the north and west, sheep are generally kept on pasture throughout the summer. In the south and south-east, especially on the Chalk and Oolite formations (semi-upland and rather poor arable), the traditional practice (now, however, becoming much less common) is to run the ewes on Down pasture, if this forms part of the holding, during only part of the day; for the rest the flock is folded on forage crops-rye, trifolium, and winter vetches in spring; clover, sainfoin, lucerne, spring vetches, mustard, and rape in summer; rape, mustard, kale, and turnips in autumn. The folding system is, of course, expensive in labour as compared with ordinary grazing, and on this account has declined very greatly during the past forty years; but it permits the keeping of a considerably larger head of stock than would otherwise be possible, particularly on the drier sort of land, where pasture is not very productive.

The stocking of pastures so as to get the best results can only be dedded by trial. The best seeds pastures will carry three ewes with, double lambs per acre; ordinary pasture perhaps two ewes (with, say, three lambs) per acre; good grassy Mil land about one per acre, and the poorer sorts of mountain land as little as one ewe to 5 or even 15 acres. On the best permanent grass it may be wise to put on no more than two ewes and their lambs to the acre and to complete the stocking with cattle; heavier stocking with sheep tends to increase the incidence of internal parasites, such as stomach worms. Most mountain grazings have a recognized carrying capacity which, however, can be improved by treatment, *e.g.* by application of lime and phosphates. In most cases, it is, of course, economically unsound to manure the whole area of the grazing, but highly beneficial results may be obtained from dressing quite small areas of the better land. The sheep are then able to adjust their mineral balance, and. make much improved growth. The direct feeding of mineral mixtures (ground , limestone, steamed bone flour, and common salt) is beneficial on hill, grazings that are very deficient in minerals. Common salt should be provided, in all cases; this can most conveniently be done by placing large lumps of rock salt under simple shelters.

Most sheep pastures are improved by the addition of a small number of cattle-say one young store bullock to every five or ten ewes. Such cattle cannot be expected to fatten, as the sheep are continually eating out the finest of the herbage; but they make quite satisfactory growth, and keep the grasses from becoming cqarseand from running to seed.

Of the various special incidents of the summer season, shearing is the first. In the south this is done as early as the beginning of May; in north-country low-ground flocks about the beginning of June, and in the highest mountain grazings as late as the first or second week of July. The date is determined mainly by the time of the natural "rise" of the wool, which becomes apparent a few weeks after the return, with the spring, of a vigorous thriving condition. The time occupied in clipping depends on the type of wool that the sheep carries. With Shortwools, less than a score will represent a day's work for a good hand-shearer, while with Blackfaces, where an inch of wool or more is left on, fully five scores per day can be shorn. In wool-producing countries like Australia, where sheep-shearing is a specialized occupation, and power-driven machines are generally used, the number clipped in relation to the quality of the work is far greater. Machine-shearing is now normal practice in the larger British flocks. Sheep may either be washed in clean stream water a week before clipping, or may be shorn in the grease. There is a loss of weight, generally from 10 to 20 per cent., if washing is done thoroughly, but the washed wool rings a proportionately higher price. Washed wools can be more accurately valued by buyers, and hence, the washing may often be profitable.

The time for summer dipping is generally about three weeks after cliping, when there is just wool enough to carry the dip. Summer dipping has the double object of killing external parasites (keds, ticks, scab-mites, lice), and of warding off attacks of blowfly. The type of dip formerly used, based usually upon sodium arsenite and sul-phur, has been replaced by a material containing usually both D.D.T. and benzene hexachloride (Gammexane) with sometimes arsenical compounds as well. The great advantage of, the new insecticides is that they give protection against blowfly, and against reinfestation by skin parasites, for a much longer period than the older materials generally for a period of about six weeks. In the use of all dips it is important to observe the directions of the makers; some shepherds tend to use stronger solutions than those recommended. Ewes and lambs are generally dipped at the same time. If it should be found necessary to dip a breeding flock, when the lambs are very young it is essential to work with very small lots of sheep, otherwise the ewe and her lamb may fail to recognize each other after the operation. Reference has already been made to dusting lambs as a means of preventing infestation with ticks.

During July and August shepherds must be continually on the outlook for sheep that have been "struck" by the fly, and the affected parts must be dressed as early as possible. The most successful preventive treatment so far discovered is dipping or spraying with a preparation of D.D.T. Spraying, as compared with dipping, makes for great economies both in time and materials. It requires special apparatus which, however, is easily portable, so that the operation in small flocks might be done on contract. Apart from preventive dipping or spraying, the clipping off of locks of dirty wool from about the hind quarters is important, as the fly is attracted by smell. Dip that is heavily contaminated with freces is worse than useless as a fly deterrent, so that the bath must be frequently emptied and refilled with clean material.

Lambs on pasture are very liable to infestation with a variety of stomach-worms and it has been a routine practice, in many flocks, to drench them with a weak solution (1 per cent) of copper sulphate. This is effective against the commonest species *(Haemonchus contortus),* but in other cases it has been necessary in the past to use either a drench of copper and nicotine sulphates or arsenical preparations. The drug phenbthiazine is very effective against most of the common stomach and intestinal worms except the very common twisted stomach-worm *(Haemonchus).* Where lambs show symptoms of heavy worm infestation-poor general condition, scouring, and anremia-it is well to have the droppings examined in order to determine the species of worms concerned. The most appropriate treatment will depend on the prevalent species. One method of administering phenothiazine is to make up a mineral lick with about 7 per cent, of the material, so that the sheep may dose themselves.

Weaning is generally done when the lambs are from three to tour months old, *i.e.* between May and August, according to the date of lambing. At weaning the ewes are put out to the poorest and barest pastures, in order that the flow of milk may be dried off, while the lambs are given the best and cleanest grass available, fields having generally been rested for the purpose; or they are folded on fresh forage crops. Clover aftermaths are ideal for the purpose, only in this case the herbage must be dry, when the lambs are turned on, otherwise many of them may get haven (bloat). Most store lambs are disposed off by the breeders at large sales which commence June, July, or August, according to the district; at the same time, or generally a little later, the breeder buys in ewe lambs to replenish his breeding flock, unless a proportion of the home-bred lambs is being retained for the purpose.

Draft and cull ewes are disposed of at other special sales in July, August, or September. On most arable farms, and on the better mountain grazings, ewes are kept in the breeding flock until they are of a certain definite age. This is generally four and a half years (four-shear or three-crop), but sometimes a year more. Oil the poorer mountain grazings, where sheep have a special value when acclimatized[1] and accustomed to the ground, there is no definite age for drafting, but each ewe is retained in the flock so long as she remains active and healthy. Even in flocks where ewes are drafted out at a fixed age, the younger ewes must be looked over and unhealthy animals, bad milkers, and ewes with defective teeth or udders culled out. Draft ewes are brought into good pasture, or folded on forage crops, as soon as the flow of milk lias ceased, and are brought forward rapidly so that they may be in good condition for early tupping. Generally, the draft lot is examined and culled, so that the major part may be sold as fully guaranteed, *i.e.* certified correct in udders and mouths. A second lot will be made of those that are broken-mouthed but otherwise correct, while unhealthy animals or those with defective. udders, etc., are sold separately for fattening.

Hill ewes after weaning and until, say, October, are herded on to the higher parts of the grazing in order to conserve the lower areas for winter use. Lambs that are not sold at weaning time are allowed to rejoin the flock after about ten days, by which time the ewes flow of milk will have ceased.

In autumn, sheep must be kept off marshy ground, water meadows, etc., as otherwise they are liable to pick up large numbers of liver-fluke larvae. When it seems likely that infestation has occurred, the sheep should be treated two or three times, at monthly intervals, commencing in September with male fern or ethyl hexachloride. On wet grazings this may be a matter of routine. Winter dipping is done during spells of good

[1] There is good reason for the belief that the main factor in acclimization is that the stock has been mildly attacked by (and has built up immunity from) louping-ill and tick-borne fever. Immunity to the former disease can now be conferred artificially. As mentioned on page 184, both diseases are carried by the tick *lxodes ricinus*, which also causes septic infections (pyremia) in young lambs.

weather in autumn, generally in October, but either before the commencement or after the close of the tupping season.[1]

Besides skin parasites, maggots, and liver-flukes, all mentioned above, there are many diseases and parasites of sheep that must be guarded against. Foot-rot is common in most breeds when they are kept on wet land. The ordinary preventive is to walk the flock through a foot-bath containing 1 ½ in. of a disinfectant and astringent solution, any actual cases having been treated in advance. Recent work suggests that the usual "walk-through" foot-bath is not very effective. It is better to place the solution in a relatively large shallow bath and let each batch of sheep stand in it for perhaps half an hour. A suitable material is a 2 per cent, solution of white arsenic dissolved by boiling with alkali. An alternative material with better powers of adhesion is 4 per cent creosote in crude oil The solution should not touch the skin; and sheep that fallen the bath must be carefully washed. The foot-bath must be covered and padlocked, when not in use.

Garget or udder clap, inflammation of the womb after lambing, scour in lambs, lung-worms and tape-worms, lamb dysentery (which can now be prevented either by vaccinating the ewes or by injecting the lambs with serum) sturdy, loupingill or trembling and braxy (which can be prevented by vaccination), scrape, tick-borne fever, pulpy kidney, and several other troubles are often responsible for serious loss; In each case the flock master must endeavour to avoid such loss by general good management as well as by special preventive measures.

Sheep often get all the water they require from plant sap, rain, and dew, and under such circumstances do not drink; but in spells of dry weather they do, so freely, and water should be offered in all cases of doubt.

Ewe lambs for breeding may be selected from home-bred stock if this be kept pure; indeed, on the poorer mountain grazings, home-bred sheep are much more likely to survive than others introduced. On the other hand, breeders who -produce cross-bred lambs for the butcher

---

[1] In certain parts of the country, there are Government regulations requiring dipping to be done at least twice a year, and between certain specified dates. Sheep moved from an area which is scheduled for scab into a " clean" area are required to be dipped twice, the second dipping being between, eight and fourteen days after the first. There were good grounds for hope, at the time of writing, that sheep-scap had been completely eradicated.

should obtain ewe lambs from others who specialize in the breeding of such hi either case the animals should be selected carefully and individually, and the number bought or retained may well be 10 or 20 per cent, greater than actual requirements, in order to allow for subsequent culling. Hoggs intended for the breeding flock should be so wintered as to keep them in healthy growing condition, but should not be made fat. Those from mountain grazings have generally to be wintered on low ground, and those from extensive mountain districts like the Scottish Highlands have sometimes to travel a hundred miles or more for wintering. Such are, almost invariably, wintered on grass alone, even if this means tot they become rather reduced in condition. Root feeding is undesirable because after such artificial treatment the sheep do not take kindly to their native hillsides. Ewe hoggs on arable or semi-arable farms may run on grass until well on in the autumn, when they are folded on roots for a gradually increasing period of the day. Later on, in January, or earlier or later according to the time of commencement of root feeding, the temporary incisor teeth begin to break off; so that roots must thereafter be fingered and fed in boxes. An allowance of 16 Lb. of roots (hardy yellow tumips or swedes) and 3/4 Lb. of hay will suffice for those of ordinary size (about 100 Lb. live weight). If such materials are not available in sufficient quantities, the bulkier concentrates such as dried grains, sugar beet pulp, and oats may be used, mixed with a little chaffed oat straw or pea haulm As soon as grass is available the hoggs will go out on the poorer and higher fields, the better pastures being, of course, reserved for the ewes. They will be shorn a week or ten days before the milking ewes, because the wool "rises" earlier. They join the ewe flock generally some weeks before the mating season, or when the old ewes are drafted out. In some flocks the gimmers (theaves) are not put to the ram until two or three weeks after the ewes, the object being to secure better weather and food conditions at their lambing time. In some mountain flocks the gimmers as well as the hoggs are "wintered away".

Fattening : Lambs of the earlier-maturing breeds can be brought to slaughter weight and condition as early as ten weeks of age. Under conditions of free markets the most popular carcass weights are probably about 25 Lb. for the small hill breeds, and 35 to 40 Lb. for others. In general (apart from cast ewes), the great majority of animals reach the meat market at ages ranging from three to fifteen months, though occasional lots of two-years-old and even three-years-old wedders of mountain breeds are still seen.

The earliest fat lambs are usually the produce of draft ewes the latter are purchased by farmers of the better arable land between July and September. Either fully guaranteed ewes or broken-mouthed ones (if they are young) may be bought-the latter being generally cheaper but requiring to have their roots cut during winter. Some are sold already tupped; others will generally be put to the ram without delay. They must have a considerable amount of hand feeding during winter, and concentrates must continue to be offered until the arrival of the "spring flush" of grass. As soon as the lambs will eat they are fed with concentrate, either along with their dams or separately. Separate feeding may be arranged by placing the lamb troughs in an enclosure with "lamb creeps" in the surrounding fence. The "creeps" may consist of special hurdles with apertures that can be adjusted to the size of the lambs and with wooden, rollers on the top and sides to avoid any risk of the lambs being caught. The "Creep Feeder" a covered trough and creep made as a unit-is even more convenient One drawback of any " creep" arrangement is that a small proportion of the lambs may tail to eat concentrated food, even if this be made attractive by the addition of locust bean or spice. An alternative arrangement is to separate the ewes and lambs at each feeding time by means of a suitable run-way and with the help of a dog. The lambs may at first be left in their enclosure, with food before them, for some time. It is then rare for any to refuse the food offered in the trough.

Traditional lamb foods are complex mixtures containing, such materials as linseed cake, bran, broken peas, kibbled or flaked maize, oats, etc., with broken locust-beans to make the mixture palatable. Compounds in the form of 'lamb nuts" are, however, more convenient and less wasteful. If folding is practised the lambs are run before the-ewes, i.e. they are allowed access to the standing crop by means of hurdles through which the ewes cannot pass, the latter following onto clean up. As soon as the lambs are fat (the ewes will generally be fat enough) they are - marketed together with their dams. Early lambs of this sort, weighing perhaps 70 to 80 Lb. live weight or 35 to 40 Lb. carcass, command a high price; but the ewe will often bring less than she cost the previous autumn. Shropshire Southdown, Suffolk, and Oxford rams are perhaps those most widely used for the production of such lambs, but Leicesters and Border Leicesters give good results with Down ewes.

The produce of draft ewes is succeeded on the market by selected single lambs from regular breeding flocks on the better arable farms, and by July and August lambs of many breeds, not excluding pure-bred Blackfaced Mountain, will be found on the fat markets. In all such cases there will have been no store period-the processes of rearing and fattening having been combined. Later on, there come lambs that have been reared as stores, weaned, very frequently sold by breeder to feeder, and finished off by the latter. The most forward of these, weighing perhaps 80 lb. or more at four months old, will be fattened, in a period of a few weeks, on clover aftermaths, rape, or other forage crops, receiving about 1 lb. per day of nutritious concentrate; they make rapid gains, as high as 4 or 5 lb. weekly. Hampshires, Oxfords, and crosses of these with Longwool and Half-bred ewes, are the favourite types for this system. Less forward lambs, including those of many mountain breeds, may be very well finished on rape or rape and rye-grass mixtures.

Lambs for "hogging" are carried on as stores, first on grass or forage crops and later on roots, until they are considered sufficiently grown to be fattened. Their general treatment meanwhile will be the same as that of ewe tegs for breeding. In most markets, under normal peace conditions, a fat live weight of not more than 120 or 130 lb. should be aimed at, even if the sheep are capable of reaching far greater size. Heavy sheep give joints that are too large for modern taste, and bring a lower price per pound. Moreover, live-weight increase becomes more expensive as the limit of growth is approached. There is also a desirable pitch of fatness beyond which it is easy to go with the fatter breeds like Oxfords and Leicesters. Lambs and young tegs are rarely too fat-the difficulty is usually to get them fat enough-but large fat carcasses, especially during spring and summer, are generally in poor demand. As a rule, therefore, tegs are kept as stores, and as cheaply as possible, until they weigh from 60 to 90 lb., according to breed, and are then put up for fattening-the latter process occupying from two to four months. The following are the examples of winter fattening rations, based on a live weight of 100 lb. (which is actually a rough average weight for fattening tegs), and each supplying about 2 3/4 lb. of dry matter, about 1.6 lb. starch equivalent, and 0.25 lb. protein equivalent. On such rations a live-weight increase of about 1 lb. per day, or rather less than 2 lb. per week, may be expected. It appears that the most uniformly satisfactory results are obtained on rations with this sort of balance, *i.e.* a ratio between protein equivalent and starch equivalent of about 1 : 6 or 7. Probably it is desirable to commence with the narrower ratio, because in the earlier part of the winter the sheep will still be in fairly rapid growth. Later, the ratio may be widened to 1 : 8 without harm.

	Dry Matter	Starch Equivalent	Protein Equivalent
14.0 lb swedes	1.61	0.98	0.098
0.5 lb clover hay	0.42	0.16	0.035
0.25 lb distillers dried grains	0.23	0.14	0.048
0.25 lb Egyptain cotton cake	0.22	0.11	0.042
0.25 lb flaked maize	0.22	0.21	0.022
Total	**2.70**	**1.60**	**0.245**

	Dry Matter	Starch Equivalent	Protein Equivalent
16.0 lb soft yellow turnips	1.36	0.80	0.064
0.75 lb seeds hay (clover)	0.63	0.24	0.045
0.50 lb flaked maize	0.44	0.42	0.043
0.25 lb decorticated earth-nut cake	0.22	0.18	0.102
Totals	**2.65**	**1.64**	**0.254**

	Dry Matter	Starch Equivalent	Protein Equivalent
10.0 lb. mangolds	1.32	0.70	0.040
0.75 lb clover hay	0.63	0.24	0.052
0.50 lb oats	0.43	0.30	0.038
0.50 lb linseed cake	0.44	0.38	0.125
Totals	**2.82**	**1.62**	**0.255**

Larger gains, up to *2.5* or 3 Lb. per week, may be obtained with the larger breeds of sheep by reducing the amount of hay to ¼ Lb. and giving a correspondingly larger allowance of concentrates, In some districts, hay is omitted altogether without any apparent detriment to the health of the sheep.

Lambs of mountain breeds, and crosses, such as Mashams and Mules (Greyfaces), may be wintered on grassland with little or no hand feeding and fattened on grass during the ensuing spring and early summer. This, naturally, is most successful when no breeding flock is maintained, the pastures being stocked in late summer with cattle only; the object being to leave a good " bite" for the winter and to avoid trouble from internal parasites. The rate of stocking in winter must, of course, be fairly low. The same practice is feasible with lowland breeds and crosses, but the sheep, at perhaps fifteen months old, are much beyond the size desired by consumers.

Apart from questions of available supplies of feeding stuffs, the following points require attention in rationing: In the early part of the winter feeding season-October and November-roots are liable to cause scouring and other disorders. The change from grass or summer forage crops to root feeding should, therefore, be made gradually, and the root ration should at first be limited. Again, if wet weather occurs and the sheep on the root land become obviously uncomfortable, they should be returned to pasture and have the roots conveyed to them. Thirdly, a proportion of the roots should be stored so that there may be no need to feed them in a frozen condition.

The choice of hay is important with feeding sheep. Pure legume hay-red clover or lucerne, for example-is usually freely eaten and is valuable on account of its high protein and mineral content. Even with this, however, unless it has been cut at an early stage of growth, fattening sheep will tend to reject the stemmy portions, and a certain amount of wastage must be permitted, if the sheep are to make rapid progress. When mixed hay is fed the animals tend to pick out the clovers and leave the grasses. Grassy hay-meadow or rye-grass, especially if it has been over-ripe when cut, is not liked by sheep, and it is frequently difficult to get fattening tegs to consume more than ¼ to ½ lb. unless by severely restricting the other ingredients in their diet. This, in turn, naturally interferes with the rate of progress.

Feeding sheep are occasionally housed in cool airy sheds. Housing may result in a substantially improved live-weight increase, if it happens that outdoor conditions are bad. But in dry weather, even if it be very cold, outdoor sheep do quite as well. The common practice on light soils is to fold the sheep on the root land, moving the fold forward as the crop is consumed so that the ground may be evenly manured. When large numbers are being fattened they are best divided into lots of not more than about 200. A man can tend about 250 to 300 fattening tegs where roots have to be cut, and about 100 more, if he has help to move the nets, feeding boxes, etc. A small portable engine for driving the root cutter effects a great saving of labour.

In spring and summer, hoggs or wedders of the later-maturing breeds are fattened on grass or forage crops, with or without the addition of concentrates. Fat sheep may be clipped as much as a month before the normal time for breeding stock.

Hoggs in wool-but with the wool dry and clean, and in fair marketable condition-may yield from 48 to 52 per cent, of their live weight as dressed carcass, the average of market sheep being nearer, the former figure ton the latter. Newly shorn sheep give about 5 per cent. more. Heavy and very fat wethers may dress upto 65 per cent, or occasionally more.

GOATS

Goats are valued either for their fleeces or their milk. In the first class are the Cashmere and the Angora, of which the latter is by far the more important commercially.

The Angora is native to the province of that name in Asia Minor. It is of a whole white colour, small in size (60 to 100 Lb. live weight), and horned in both sexes. The fleece (mohair) varies a good deal in quality, but in the best specimens is extremely tine, silky, and lustrous. Typically it is from 8 to 10 in. long and hangs in separate and well-defined ringlets. Four pounds is an average weight of fleece, but superior specimens shear 6 lb. and occasionally upto 14 lb. Fine mohairs have a distinct use in the manufacture of plushes, velvets, and bright lustre materials, and .command a price comparable to that of fine wool. The quantity used annually in the Bradford district runs into thousands of tons. The flesh of the young and fat Angora is said to be good. The breed is not very prolific, twins being rare and the annual increase being generally under 80 per cent. Angoras require a dry climate but are otherwise hardy, and thrive under semi-arid conditions where sheep do not prove a complete success. The chief mohair-producing countries in order of importance are South Africa, Asia Minor, and the United States (western and south-western states).

Milk goats are of several breeds; the best known are the Toggenburg and Saanen (Swiss), the Nubian, and the Maltese. In Britain, we have the imported Swiss and Nubian types, as well as somewhat ill-defined sorts like the Anglo-Swiss, Anglo-Nubian, and Anglo-Nubian-Swiss derived from crosses of the foregoing with the native type. Goats milk is comparable in composition with cows-somewhat more variable though probably, on the average, rather richer in fat and poorer in other solids. It is often more easily digested by delicate persons, and has the advantage that it can be reckoned to be free from tubercle bacilli, which are the commonest of the more dangerous organisms found in cows milk. Does of good type reach yields of 3 to 5 qt. per day and occasionally upto 2 gals. The lactation period is ordinarily five or six months, but may extend to nine or ten. Lactation yields of over 400 gals, have been recorded in ten months, and the average in good recorded herds is over 200 gals. The highest recorded lactation yield up till 1946 was of 5928 Lb. of milk in 338 days. Goats, as is well known, will live on almost any sort of herbage, and in mountainous districts, or wherever food is scanty, they form a valuable source of milk. It seems unlikely, however, that they will ever become commercially important in lowland districts, where improved dairy cattle can be kept.

PLATE 20

A. THOROUGHBRED STALLION

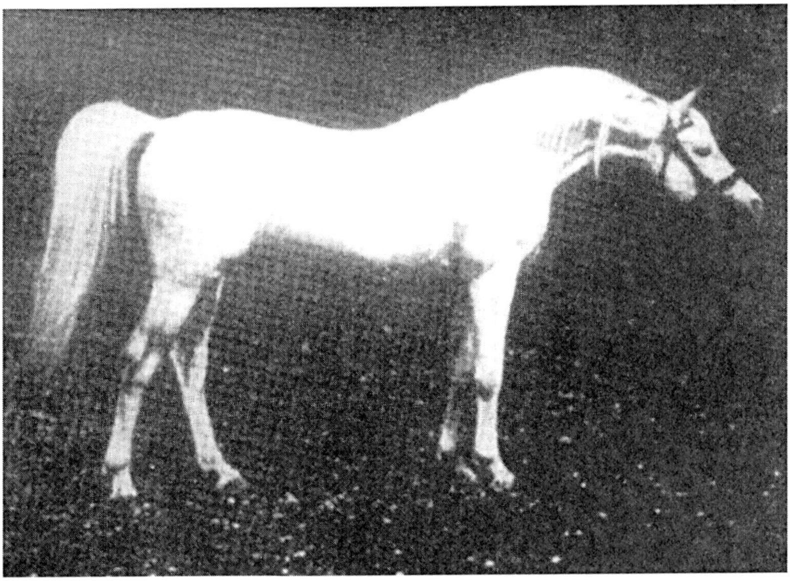

B. ARAB STALLION

A. HUNTER GELDING

B. POLO PONY MARE AND FOAL

CHAPTER VI

HORSES

THE horse family (Equidae) forms part of the group of Perissodactyla, or odd-toed ungulates. Zoologists include all members of the family, whether horses, asses, or zebras, under the genus Equus.

The Zebras, which are found only in Africa, include the now extinct Quagga (Cape Colony), the True or Mountain Zebra (south-western Africa), Burchell's Zebra (Transvaal, etc.), Grevy's Zebra (Somaliland), and two or three more species or sub-species related to one or other of the foregoing. None of the zebras has attained any importance as a domesticated animal, although specimens of several species have been successfully broken both to harness and the saddle. Grevy's Zebra has generally been regarded as the most promising from the utility point of view.

The Wild Asses embrace two African types, the Somaliland and the Nubian; and two Asiatic, the Onager of Syria, Persia, Arabia, and India, and the Kiang of Turkestan, Mongolia, and Tibet. Domesticated donkeys are generally regarded as of purely African origin, but some authors have supposed that Onager blood has been blended with African in some of the larger breeds.

The only truly wild horse now existing is the *Equus Przevalskii* of the Gobi Desert and north-western Mongolia. Whether or not this wild species has been concerned in the origin of our domesticated horses is somewhat uncertain; that it is not the sole ancestor is fairly generally agreed, most authorities supposing that at least three distinct species are represented among our modern breeds. Although these original species have been crossed and intermingled to a considerable extent, we may still distinguish fairly pure representatives in the Celtic pony *(Equus celticus)* of the Western Isles; the Arab, Barb, etc. (Oriental light-legged type), and the Shire and Belgian (Western heavy-legged type).

Hybrids between horses and zebras, asses and zebras, or horses and asses maybe produced with comparative ease. The hybrids, however, with rare and doubtful exceptions, are sterile. Of the hybrids the only one of much economic importance is that between the male donkey or jackass and the mare, viz., the mule.

The opposite cross, between a stallion and a she ass, is of much less value on account of its smaller size.

As a preliminary to a description of the various types of horses some explanation of "points" and gaits is necessary. The points are shown and named in Fig. 13. of the things that go to the making of a good horse, some are related to the special purpose for which the particular animal is to be used, *e.g.* draft, harness, or saddle. Other things are regarded as desirable in all breeds and types, and these may be now briefly mentioned.

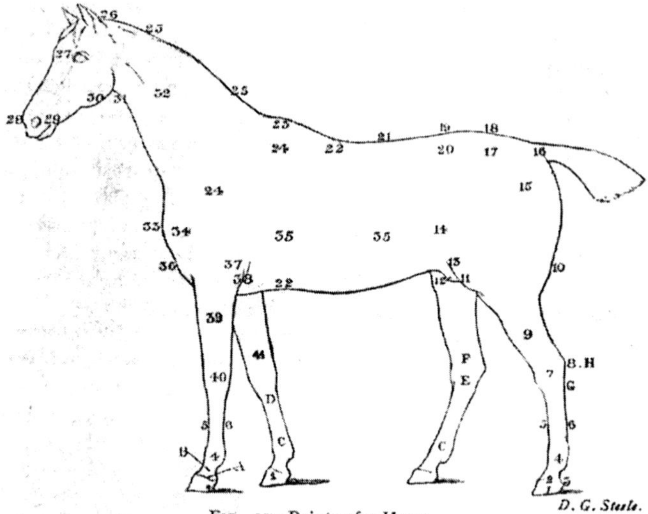

FIG. 39—Points of a Horse

1. Hoof
2. Coronet
3. Heel
4. Fetlock or pastern joint.
   B. The pastern
5. Cannon bone
6. Back sinew or tendon
7. Hock
8. Point of hock
9. Second thigh or gaskin
10. Haunch or lower buttock
10–20. Thigh
11. Stifle
12. Sheath
13–14. Flank
15. Hip joint
16. Root of tail or dock
17. Rump
18. Croup or "quarter"
19. Loins
20. Point of hip bone
21. Back
22–22. Girth or chest measurement
23. Withers
24–24. Shoulder blade (*scapula*)
25–25. Crest
26. Poll
27. Forehead
28. Muzzle
29. Nostril
30. Jaw
31. Throat or windpipe
32. Neck
33. Point of shoulder
34. Shoulder
35–35. Front ribs and short ribs behind, forming the barrel
36. Chest or breast
37–34. True arm (*humerus*)
38. Elbow
39. Arm (or fore arm)
40. Knee
41. Chestnut, castor or callosity

Seats of Common Diseases.

A. Side-bone
B. Ring-bone
C. Wind-gall
D. Splint
E. Spavin
F. Thorough-pin
G. Curb
H. Capped hock

FIG. 13 : Points of Horses

The head should be of good width between the eyes, and the eyes themselves prominent and of good size. These points denote intelligence and docility. The nostrils should be large and open to permit of free breathing. The neck should be of good length. The shoulder should be at least moderately sloped back, as an upright scapula almost invariably makes for a stilted gait. The back and loin should be short, nearly straight, and thickly clad with muscle in order to give strength; too short a back, however, is often associated with a lack of freedom in movement, especially at the trot. The hind quarters should be long and wide, and not too steeply sloping to the tail. Large fore arms and thighs are important as indicating strength. Chest and middle should be deep in order to give, on the one hand, space for heart and lungs, and on the other, capacity for food. For similar reasons the ribs should be well "sprung" or rounded. The knee should be large, the hock broad, and the cannon bones "flat," *i.e.* the tendons should be large and set well back from the bone itself, in order to give the necessary strength and freedom of movement. A fair length and moderate slope of pastern are necessary in order to break the concussion when the foot meets the ground, but too much length or slope involves weakness. The feet should be wide in proportion to their length, the soles concave, and the substance of the hoof tough and durable. The legs should be well knit under the body, and should be moved freely and straight.

The gaits natural to most horses are the *walk, trot,* and *gallop.* Some horses develop, also naturally, either the *amble* (or running walk) or the *pace.* The chief additional gaits to which horses may be trained are the *canter,* which is a restrained and modified gallop, and the *rack* or "single-foot." Of these various modes of locomotion the simplest are the trot and pace. The trot is a diagonal two-beat gait-that is, the off fore and near hind move together and meet the ground at the same time, making one beat; the near fore and off hind, similarly move together and produce the second beat, the intervals between the beats being the same. Young horses frequently "strike" or "click" when at the trot, the toe of the hind shoe meeting the sole of the fore shoe just as the former is meeting and the latter leaving the ground. In the trot the rider generally "posts"; that is, he rises in his stirrups in unison with one or other pair of legs-generally with the near hind and off fore-and thus, escapes the jolt of the second beat.

In the pace (which is rarely found except in American pacer-bred animals) the left pair and the right pair of legs move together, as in a camel's "trot." This sidelong gait is both uncomfortable and unsafe for a rider, but in a harness animal it is not objectionable, and is on the average slightly faster than the trot.

The walk, amble, and rack may be grouped as symmetrical four-beat gaits-that is, each foot is moved in different time from the others and makes a distinct beat, while the movement of the left leg, in each case, is similar in amount and opposite in time to that of its right fellow. The chief difference between the three is in speed. The amble is performed at a rate intermediate between those of the walk and trot. It is a good saddle gait, comfortable for the rider and easy for the horse, and is favoured in many countries where long distances have to be travelled. The rack is performed at a speed faster than that of a sharp trot. It is a showy gait, and a comfortable one for the rider, who has nothing to do but to sit still in the saddle. It makes, however, very heavy work for the horse; animals are trained to it with difficulty, and it is regularly developed only in the American saddle breed.

The canter and gallop are unsymmetrical gaits, *i.e.* the movements of the one leg are not matched by those of its fellow. A horse may canter or gallop either "right leg leading" or "left leg leading," and may be trained to strike off in either way or to "change the leg" at the rider's will. In the gallop the four beats follow each other in very rapid succession, and there succeeds a relatively long interval in which each leg in turn leaves the ground, and the horse is suspended in mid-air with all four gathered under him. The canter is a very comfortable gait for the rider, and in saddle-bred animals it forms an easy one for the horse.

The particular gaits that are required vary with the purpose for which the animal is to be used. Heavy draft animals perform all their work at a walk, but should be able to trot freely and easily on occasion. The principal gait of the harness animal is the trot (or alternatively and exceptionally, the pace), but a free and rapid walk is also to be desired. Ordinary saddle horses should be trained to walk, trot canter, or gallop at the rider's will, while the American " gaited" saddle horse is required to show five distinct gaits, viz., the walk, trot, canter, and rack, together with the walking trot, or one or other of two somewhat similar slow movements.

From the utility point of view, horses (including ponies) are classified as saddle, harness, draft, and pack. The first type is required to carry a rider, the second to draw a light vehicle at considerable speed, the third to pull, and the last to carry loads at a moderate or slow speed.

The Saddle Horse : As far as concerns this country, the hunter and the polo pony may be taken as representative types of saddle horses. The former is typical of the larger, the latter of the small sort of animal for actual utility purposes.

Hunters are classified according to the weight they are judged to be capable of carrying in the field. One system is to make 12 st. the minimum for the light-weight class, and 13 st 7 Lb. the minimum for the heavy. More elaborate classifications are adopted at the larger shows. The qualities specially desired in a Hunter may be summarised as follows:

1. *Speed :* He must be fast at the gallop and should show a collected and easy movement at the walk, trot, and canter. Very high action is generally incompatible with great speed, and hence is not wanted.
2. *Staying Power :* He should be able to gallop at a good speed over three or four miles of ordinary mixed country, taking fences as he goes. Staying power can never be finally judged without actual test, but a large nostril, a good width between the angles of the jaw, and a deep capacious chest are useful indications of it.
3. *Strength :* The muscles chiefly used in galloping under a rider are those of the back, loin, and hind quarter. The back and loin should be short, wide, and thickly covered with muscle, while the quarter should be long, wide, and strong, the muscles carried well down towards the hocks, and the gaskin or second thigh wide from front to back.
4. *Durability :* The limbs and joints must be strong enough to withstand the stresses to which they are subjected. The heaviest of these occur during jumping-on rising they are greatest in the hock, and on landing in the tendons and ligaments of the fore legs and the bones of the pasterns. Hence, broad clean hocks-neither too straight nor too bent-and large bone and tendons in the fore leg are points of great importance.
5. *Temperament :* The animal must have plenty of courage and yet be susceptible to such training as will place it completely under the control of its rider. This may be judged to some extent by the

appearance of the head, the expression of the eye, and the carriage of the head and ears. A long neck, fine at its junction with the head, generally makes for easy control by the bit, whereas a short, thick neck generally means difficulty in control. Jumping ability, which is largely a matter of temperament, cannot be judged except by trial.

6. *The Comfort of the Rider : The* withers should be high and fairly thin, to provide a comfortable seat. The gaits should be smooth and easy and the stride long, the feet being carried rather low, especially at the canter and gallop. Good length and slope of pastern also make for comfort by breaking the impact of the leg on the ground.

In size, hunters range from about 15.1 to nearly 17 hands,[1] most being between 15.3 and 16.2. The weight-carrying capacity is not a function of height but rather of general build, the shorter-legged and stronger-boned animals being generally upto greater weights, if somewhat slower, or the taller and finer-boned types.

The polo pony may best be described as a weight-carrying hunter in miniature, with, if possible, a greater measure of intelligence, in order to permit of greater refinements in training. Great nimbleness and particularly cleverness in turning are necessary in playing ponies.

The American or gaited saddle horse is an interesting type that lias been developed in the direction of great style and symmetry, combined with high endurance at moderate speed, and a great measure of comfort for the rider. It is, however, inferior to the hunter in strength, speed, and fencing ability.

Harness Horses are now bred in very small numbers and the interested reader must be referred to other works for information about them.

Draft Horses : Of chief importance to agriculture is the draft type, more especially the heavy draft; the qualities desirable in the latter may be summarized as follows :

1. *Strength : This* is to some extent proportional to weight, but depends also in large measure on the development of particular muscles and on the efficiency of the transmission of the available power

[1] The height of horses is measured to the withers, and the unit used is the hand = 4 in.

through the limbs. The back and loin should be short, wide, and thickly covered with muscle, while the quarters should be long, wide, and well muscled down to the thighs. Arms and gaskins should be broad, with the muscles prominent. As far as concerns the transmission of power, the legs should be closely knit to the body and placed well underneath it; they should be moved freely and straight Serious defects under this head would be a long and bare back, narrow quarters and thin upper thighs, wide waddling action in front, or a tendency for the hocks to twist outwards, when they are submitted to the strain of draft.

2. *Endurance : Lung* space-a deep, capacious chest-is of importance, but perhaps of rather less consequence in the draft animal than in horses tot work at a high speed. Most horses that fail to stand up to severe and long-continued work in draft do so from an insufficient capacity for food. The good feeder, who can be known by his well-rounded ribs and deep capacious belly, will generally be found to have great endurance, whereas the tall, light-ribbed animal with flat sides readily gets weak and out of condition.

3. *Wearing Qualities :* Apart from unsoundnesses like roaring and stringhalt, the useful life of the draft animal is terminated, in a large majority of cases, by some breakdown in the feet and legs. The common troubles are sprain of the tendons, ring-bone, side-bone, and navicular or other disease of the foot. Hence, special attention must be paid to " under-pinning." The cannon bones should be of good size, but must show "quality," *i.e.* they should be clean and hard, with the tendons well set back and large, giving the leg a markedly flat shape. The pasterns should be of good length (they can, however, be too long) and should slope at about 45°. The foot should be wide at the hoof head, neither flat nor steep, and the substance of the hoof should be tough and strong.

4. *Speed : The* draft animal should have a long stride and a free, straight, active, and easy walk. An animal with a short stepping, stilted action either cannot walk out at a reasonable speed, or does so with a great wastage of effort. The animal should be able to trot freely and rapidly when required, but mere is no particular advantage in high knee or hock action.

5. *Temperament :* The draft horse should have pluck and energy, without being highly strung or intractable. His head will show less refinement, but more strength than tot of the saddle type; it should be carried fairly high and the expression should be alert.

As regards size, it may be said that a height of 17 hands is sufficient for a heavy draft animal, if he be thick of body and proportionately built. Many good horses are under this height, and females are generally considered big enough for breeding, if they reach 16.2. A good weight for a mature gelding in working condition is 14 to 15 cwt, or, say, 1600 lb. The largest stallions in show condition occasionally exceed a ton (2240 lb.), though 18 cwt. or 2000 lb. is more usual.

The light draft type may be described as something intermediate between a heavy draft and either a weight-carrying hunter or a harness horse. It will show less weight and strength than the heavy draft, but more activity and speed: Since it will have to perform a good deal of its work at the trot, it should show freedom and ease in this gait, and should be hard-limbed and sound of feet to withstand the heavy wear and tear of fast street work. In general, the light draft type is not a very profitable one to breed, since the market requirements for vanners and light lorry animals are now very small. It makes, however, a useful farm horse for the lighter classes of soils, and on certain farms for miscellaneous light tasks.

The pack type of horse is one of minor and decreasing importance, since its use is confined to districts, where the roads are not fit to carry vehicular traffic. The most generally useful sort of animal for the purpose is a large pony or small sturdily built horse, sure-footed, active, and docile.

BRITISH BREEDS

The Thoroughbred or English racehorse (Plate 20, *A)* is of almost pure Eastern blood. It may be said to have originated in an importation of some forty mares-mostly Barbs-made by order of Charles II. During Charles's lifetime these mares were kept as a Royal Stud, and at his death they were sold and dispersed throughout the country. Of many stallions imported both before and after that time, by much the most important were the Byerly Turk, imported in 1689, Darley Arabian (1706), and Godolphin Barb (1724). The blending of these different blood lines, together with long-continued and rigorous selection on the basis of racecourse performance, has produced an animal that is far superior in speed to any modern Oriental breed, and is by much the fastest horse in the world.

Modern Thoroughbreds show a good deal of variation in type, chiefly because there is, and has been, little or no selection except that based on racecourse performance. The height varies from the one extreme of under 14.2 to the other of over 17 hands, the average being probably between 15.2 and 15.3. The head is lean and refined, the eye large and prominent, and the nostril wide. The courage of the breed is proverbial, but the temperament is sometimes objectionably highly strung and occasionally vicious. The neck is long, rather thin and muscular, with a large windpipe, and is carried rather low. The chest is capacious, but deep rather than wide, and the shoulder is preferably but by no means invariably well sloped. The back and loin are very generally short, straight, and muscular; the croup is commonly of good length but varies in conformation, being sometimes level, with the tail set on rather high, sometimes rather steeply drooping. The bones are clean and hard, with the tendons sharply defined and well set back. Small "weedy" bone and long slight pasterns are objectionable. The movement should be smooth and easy, the stride long, and the legs carried rather low. The commonest colours are bay and brown, with chestnut, grey, and black following in order of frequency.

From the utility point of view Thoroughbreds vary greatly in value. Natural toughness, stamina, courage and speed are constant assets, but it is only when these qualities are combined with strength and wearing qualities that they make for commercial value. The long-legged, slightly built, and light-boned animal is often a good flat racer, but otherwise is of very little use. Such " weedy" animals, if they fail to win on the racecourse, are almost worthless. On the other hand, the stoutly built and bigboned animal, even if he is not fit to win races, forms an ideal saddle-horse sire and possesses great commercial value. Many breeds of light-legged horses are indebted to the Thoroughbred for endurance, speed, and courage.

The Arab (Plate 20, *B)* has a history that stretches back to about the beginning of the Christian era, arid the purity of the breed has been jealously maintained for the greater part of this long period. Knowledge of pedigrees extends over many generations. Careful selection, with very special attention to such qualities as courage and endurance, has been carried on by the Arabs for a long period, and the modern breed is distinguished by these qualities in a very marked degree. From the Thoroughbred, to which he is closely related, and which he resembles in a general way, the Arab is distinguished by several fairly constant characters. On the average he is smaller, 14.2 hands being perhaps the average height. The head is wider between the eyes and tapers more towards the nose; the neck is more arched, the head is carried higher and the shoulder is more oblique; the croup is generally longer and typically more level, with the tail higher set on; the withers are commonly rather thicker.

In speed the Arab is far inferior to the Thoroughbred. He is, however, capable of carrying a heavy weight in proportion to his size and is unexcelled in endurance; many animals have been ridden 50 or 60 miles a day for several days on end. In long distance tests, the breed is generally supreme. An Arab Horse Society has been established in England, and published the first volume of its Stud Book in 1919.

The Hunter and the Polo Pony (Plate 21) cannot yet be regarded as breeds in the strict sense, but are rather commercial types. Registration with a view to the establishment of actual breeds lias, however, commenced, foundation stock being entered by inspection. Hunters generally carry a large proportion of Thoroughbred blood, the other elements in their breeding being very varied. "Clean bred" hunters are by no means unknown, but it is exceptional to find Thoroughbreds with the necessary bone and substance to carry heavy weights. The type required in flat country, such as the Midlands is different from that suited to hilly areas; the latter must be short-legged and sure-footed rather than specially fast. The polo-pony breed, when formed, will contain the blood of several native breeds of ponies such as the Welsh, Exmoor, and Dartmoor, blended with that of the Thoroughbred and Arab.

Ponies : Of the native British ponies, the smallest and certainly one of the oldest breeds is the **Shetland.** Standing generally under 40 in. in height (the maximum height for mature animals is 42 in.), the Shetlander is very stoutly built-rather like a heavy draft animal in miniature-and very strong for its size. In their native isles the ponies are used as pack animals, especially for the carrying of peat. Being generally docile and easily trained, they are very popular as riding ponies for young children.

The Western Isles Pony, or light-type Highland pony, is native to the Hebrides, some of the best and most typical being found on Barra. It is generally between 12 and 13 hands high and is of good saddle conformation, though often steep of shoulder and rather sickle-hocked. The quality of bone is excellent and the breed is very active and, of course, exceedingly hardy. It is used, like the Shetland, principally as a pack animal.

The heavy or Mainland type of **Highland** Pony, frequently spoken of as the Garron (Plate *22, A),* is found in the Central Highlands of Scotland, notably in Perthshire. Averaging nearly 14½ hands in height (the maximum allowed by the breed standard) the breed is very stoutly built, distinctly "drafty" in type, with rather heavy shoulders and thick withers. The neck is highly crested, the mane abundant. The" bone" is sometimes too round and the legs carry, in winter, a good deal of long feather. Distinguished for his great hardiness and longevity, the Highland pony is used by the smaller glen farmer as an all-purpose horse for riding, driving, hauling, and carrying; during the stalking and shooting seasons many find employment in carrying deer or game. Grey is perhaps the commonest colour.

Of rather similar type to the last is the **Fell Pony** of Cumberland, Westmorland, and West Yorks. It is generally some 2 or 3 in smaller than the Highland, but is built on the same general plan-thick and short of leg, with large bone. It is believed to be descended from the old Galloway pony, which was famed for its toughness and strength. The **Dales Pony** of East Yorks and Durham is similar to the Highland pony but generally bay or brown in colour. It is larger than the Fell pony and might almost be regarded as a small draft horse, since many specimens exceed 14.2 or even 15 hands in height. The general conformation is rather massive, but the breed is both sure-footed and fast. In past times many were employed for pack work, and some are still taken for colliery purposes: their chief use, however, is for farm work on the smaller holdings of the Dales.

The Exmoor and Dartmoor are ponies of saddle type. The latter stand about 12.2 hands, the former perhaps 2 or 3 in. less. They have well-bred heads, good shoulders, and hard bones, but, rather frequently, drooping quarters and sickle hocks. Crossed with small Thoroughbred or Arab sires, they produce excellent riding ponies.

The New Forest type shows rather less breeding than the foregoing, and generally a heavier and more upright shoulder; the legs, however, are better built. Forest-bred specimens run from 12.2 to 12.3 hands.

The Welsh Pony (Plate 22, *B),* which has existed from very early times on the moors and mountains of Wales, is a breed of very considerable importance, forming perhaps the most valuable foundation stock for the production of saddle ponies. Two types are recognized and are registered separately in the Stud Book. The smaller and lighter type (under 12 hands) is of saddle conformation, while the larger (upto 12 ½ hands) has higher action and is more of the harness type. In general these ponies are very hardy and enduring, and show great freedom and speed in movement.

The Welsh Cob is derived partly from the foregoing, with an infusion, in early times, of some Thoroughbred, Arab, and Hackney blood. Typical specimens measure from 13.2 to 14.2, have durable limbs, and show high-stepping action and considerable staying power. They are used as riding animals *by* farmers in the remoter parts of Wales.

The Heavy Draft Breeds now recognized in Britain are the native Shire, Clydesdale, and Suffolk, and the recently introduced Percheron and Belgian The Clydesdale is the breed of Scotland and the four northern counties of England, and is the chief of the draft breeds in Ireland. The Suffolk is native to East Anglia and has only recently spread beyond the borders of this province. The Shire is found throughout the rest of England and Wales, excepting only mountain and moorland districts.

The Shire (Plate 23) is regarded as the direct descendant of the Old English war horse, or Great Horse. This lattèr was derived partly from the horses that existed in Britain before the, Roman invasion, partly from heavy horses that were introduced at various times, chiefly from Flanders. In olden times the Great Horse was valued for his ability to carry the immense weight of armour, etc., of the mounted knight, and various kings of England from John to Henry VIll-were at considerable pains to maintain or increase the size and quality of the breed. The use of horses as the chief draft animals on the farm is a comparatively recent thing, for till the eighteenth century over most of this country the bulk of field work was done by oxen.

Careful and systematic breeding of Shires has been carried on since the middle of the eighteenth century, and some modern pedigrees go back a hundred years or more, in the days before the formation of a breed society (1878) there was a good deal of variation in type from one district to another, but the breed has been levelled up by continuous interbreeding of the local variations. Of the early Shires it may be said that many were weighty and powerful animals; but upright pasterns and steep, narrow hoofs, a want of quality in bone, and unsoundnesses like side bone, were common faults. The aim of the Shire breeders has been to eliminate these defects, while preserving or even increasing the size and weight. The strict veterinary inspection of all animals competing at the shows of the Shire Horse Society has had a very beneficial effect. Twenty years ago there was a tendency to aim at an excessive quantity of "feather," but the present-day preference is for legs almost as " clean" as a Clydesdale's.

The modern Shire is probably the weightiest of all horses, though the Belgian is sometimes regarded as first in this respect. Mature stallions of the best class, measuring generally 17 hands or over, often reach a weight of 20 cwt, when in show condition. The head is heavier and coarser than in other British breeds and the eye is smaller. Typically, the back is short, wide, and very strong; the ribs both deep and widely sprung, giving a middle of great capacity; the quarters very wide and powerful, but often somewhat short and drooping. The fore arms and thighs are of great size, and the bone should measure at least 11 in below the knee and 12 in below the hock-the larger the better, so long as it is clean and flat. The pasterns are of moderate length and preferably well-sloped. The feet are now generally of good sizewide at the heels and open at the hoof head. The "feather" is long and straight, and rather strong in quality. Curly or woolly feather is regarded as objectionable. Common colours are bay, brown, and black, with grey, chestnut, and roan less prevalent The mare, apart from her smaller size and the feminine character in her head, is relatively longer in body and shorter-legged than the stallion. The temperament of the Shire is commonly very docile-sometimes phlegmatic. In general, the animal gives the impression of great strength and massiveness. He is an excellent feeder and has great endurance on slow work.

The Shire, while fairly popular in many foreign countries, has not on the whole competed very successfully with other draft breeds.

A cross between the Shire and Clydesdale combines a good many of the best qualities. The progeny of a mare of this cross by a registered Shire stallion is eligible for registration in the Shire Stud Book.

The Clydesdale (Plate 24) may be said to have had its origin during the first half of the eighteenth century in the interbreeding of the native horses of the Upper Ward of Lanarkshire with colts introduced from England; some at least of the latter were of Flemish breeding. The most famous of the pioneer breeders was John Paterson of Lochlyoch who, about 1715, introduced from England a black Flemish stallion which has been regarded as the foundation sire of the breed. The earliest horse to which pedigrees can now be traced is "Glancer," foaled about 1810, and probably descended, in the female line, from the Lochlyoch stud. Occasional crossing with Shires was done up till about 1890, and a good many three-quarter blood Clydesdales, with the Shire cross, are entered in the earlier volumes of the Stud Book Many breeders believe that the separation of the breed from the Shire was a mistake, as the original differences were not very marked. At times the breeds have drifted very far apart, but the less massive and cleaner-legged type of Shire that is now preferred is not very different from the heavier-boned sorts of Clydesdale.

The Clydesdale is similar in height to the Shire, the stallion being 17 hands or over and the mare about 16.2. The general build, however, is less massive, the legs being proportionately longer, and the body neither so deep nor so wide. The head shows more quality and refinement, the eye being larger and the muzzle finer. The neck is rather longer and more crested, the head more gaily carried, and the shoulder has on the average more slope. The back is more frequently faulty than in the Shire, being sometimes too long, sometimes hollow, and sometimes rather bare of muscle. Flatness of rib and shallowness of heart or belly are commoner faults. The quarters are usually longer and more level, but narrower when viewed from behind. The arms and gaskins are not so large and are shorter in proportion to the length of the cannon. The bone of the Clydesdale is much smaller but is harder, flatter, and cleaner, and the feather, which is more silky in quality, is confined to the back of the leg. The pasterns are longer and have more slope; the feet are relatively larger and wider but with a more frequent tendency to flatness. The Clydesdale, as a breed, is very free from leg troubles like grease, side-bone, ring-bone, navicular disease, etc., and the average working life, particularly when tested against other breeds on paved streets, is exceptionally long. The breed is also remarkable for its straight, free, and close movement, and shows great ability and cleverness in handling loads, in general, it may be said that the Shire excels in weight and strength, and in endurance;

while in speed, agility, and wearing qualities the Clydesdale is preeminent. The temperament of the Clydesdale is more mettlesome and nervous than that of the Shire; he is more troublesome to break and requires more horsemanship on the part of his driver. Another point of difference is that the Shire matures at a much earlier age than the Clydesdale, so that the difference in weight is far more marked at the yearling or two-years-old stages than in aged animals. Clydesdales are of all the ordinary colours, bay and brown pre-dominating, while chestnut and grey are rare. White faces and legs are almost constant features, and splashes of white on the belly, or white hairs irregularly mixed throughout the coat, are becoming increasingly common.

The Clydesdale is a common draft breed in Canada, New Zealand, Australia, the United States, and many other countries. Broadly speaking, however, it is less popular than the Percheron. The chief defects of the breed, as urged, *e.g.* in the United States, are its tendency to flatness and shallowness of rib, and narrowness of build, with a consequent inability to keep in condition, when exposed to heat or cold or subjected to severe work or scarcity of food.

The Suffolk (Plate 25, *A)* is the least numerously represented of the three native draft breeds. It has been recognized as a distinct local type for several centuries, but some infusion of trotting blood was made over a hundred years ago. The average height is 1 or 2 in. less than that of the Clydesdale or Shire, but the weight is not notably less than that of the Clydesdale. The general build is thick, wide, and low set. The ears are rather short, the neck crested, the chest deep and wide, the barrel large, the ribs well-rounded, and the quarters wide and heavily muscled, though often rather steeply sloping. The bone is rather light, but not more so than in many modern Clydesdales. The legs carry practically no feather. The pasterns are rather short and upright, especially the hind ones. The feet formerly showed a tendency to flatness but the breed lias been greatly improved in this respect. The Suffolk is active, but the movement is not so straight and true as in the Clydesdale. The colour is some shade of chestnut, with sometimes a white spot or blaze on the face. White feet are regarded as objectionable. The outstanding merit of the Suffolk is his constitution. When well-fed he puts on a great reserve of that, and he is easily kept. On ordinary work he can go long hours without food, and stands exposure and scarcity very well. In draft he is remarkably steady and he lives and works to a long age. The mares are exceptionally regular breeders. The Suffolk was long in obtaining recognition outside his native district, but during recent years several studs have been established in other parts of England.

The Percheron (Plate 25, *B)* derives its name from the old French district of La Perche, lying within a radius of some forty miles from the town of Nogent-le-Rotrou, situated about a hundred miles south-west of Paris. The breed contains a good deal of the same blood as the other draft breeds, but was influenced to a considerable extent by the introduction of Eastern (Arab) blood. In coaching times, the Percheron was used chiefly for road work, and its evolution into a heavy draft breed has occurred since the widespread development of railways. Its main characteristics may be summarized as follows:

The height averages 16.2 for stallions and 16 hands for mares. The weight is rather great in proportion to the height, but even by this measure the Percheron is perhaps the smallest of the heavy draft breeds. The head is rather lean and shows a good deal of breeding. The neck is highly crested and the shoulder long and oblique, the chest deep and wide. Width and strength of back and roundness of rib are very striking characteristics, but the depth of back ribs is not so uniformly good. The quarters are wide, but often short and generally rather steeply drooping, the tail being set on low. The. bone is fairly hard and fine though somewhat round, and the legs are free from long hair. The pasterns are somewhat short and upright, and the feet of reasonable size, shapely, and very sound and durable. The hind leg has often too much "set" *-i.e.* is too crooked- to please the eye of British judges, and the hock is often lacking in breadth and is too round and fleshy. The movement is sharp, with rather high knee action at the trot, but the front action is not so straight as could be desired. The temperament is very good-docile, yet active and willing. The common colours are grey and black, with bay and brown rare. The Percheron is the favourite draft breed in the United States, where it owes its popularity to its clean legs, its quiet temperament, good constitution, and its high power of endurance on rather fast work. Importations of pure-bred Percherons into Britain were of no importance until 1918, when the British Percheron Horse Society was formed.

PLATE 22

A. HIGHLAND PONY MARE

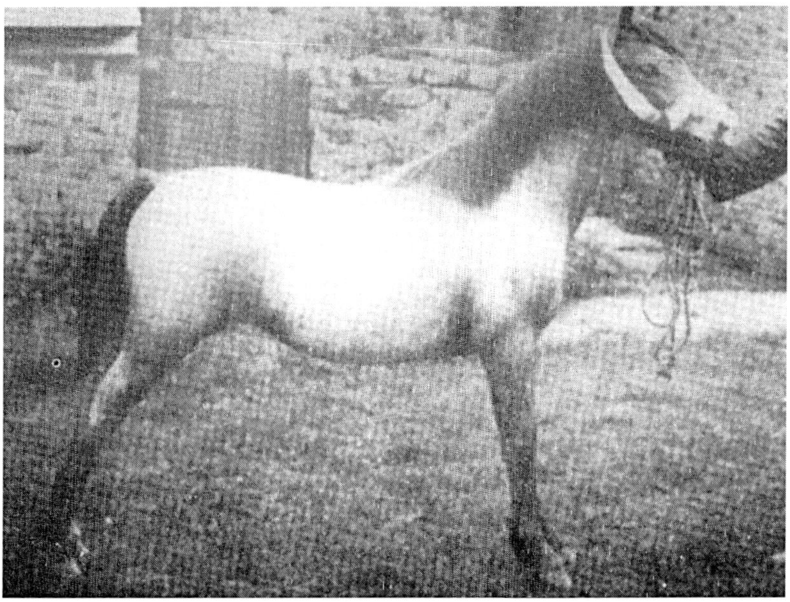

B. WELSH PONY

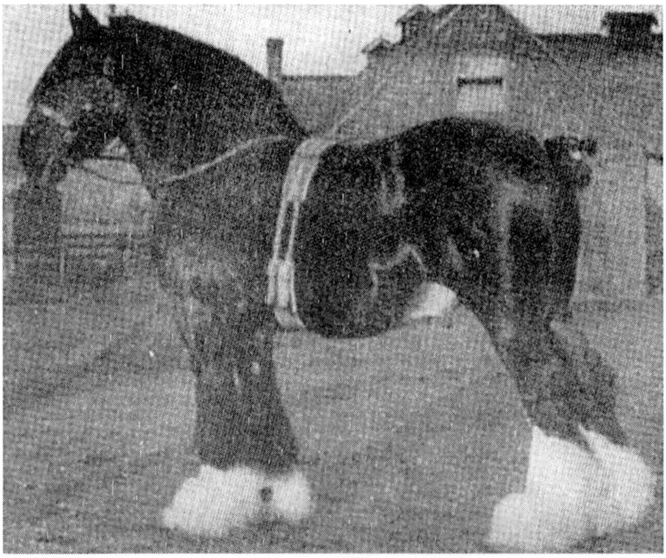

A. SHIRE STALLION

B. SHIRE MARE

PLATE 24

A. CLYDESDALE STALLION

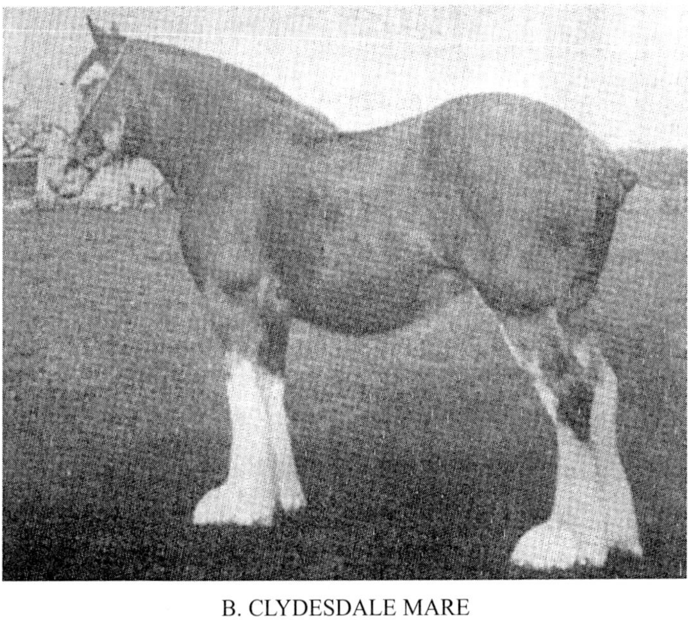

B. CLYDESDALE MARE

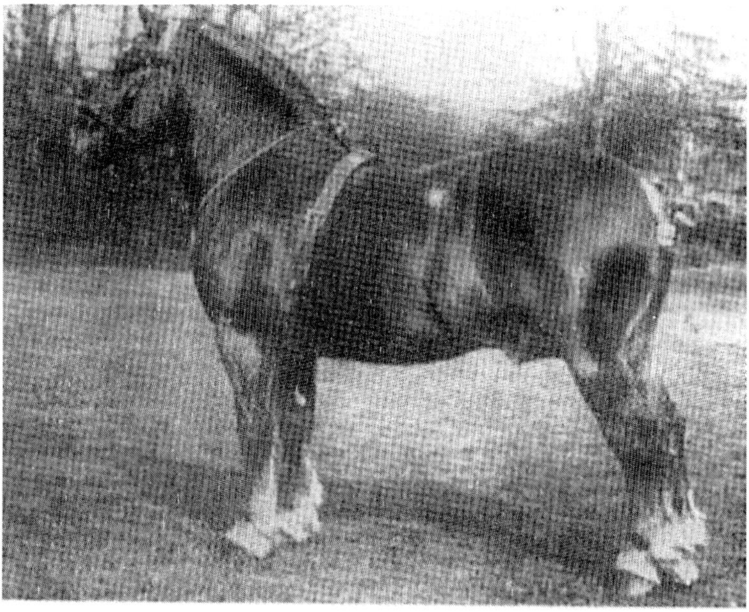

A. SUFFOLK STALLION

B. PERCHERON STALLION

Belgian horses were imported into Britain in considerable numbers during the few years up till 1939. The breed is a very old one, related to the Flemish animals which played a part in the formation of the Shire and Clydesdale breeds. The Stud Book dates from 1886. The general build of the Belgian is heavy and compact, with a very large girth in proportion to the height. The neck is short, the back straight and strong, the rump rather short and drooping. The short legs are practically devoid of feather. The bone is rather round and the feet tend to be small. The temperament is very docile and the horses are willing workers. The commonest colour is chestnut, though bay, brown, and roan also occur.

BREEDS AND NUMBERS OF LICENSED STALLIONS

The relative numerical importance of our heavy draft breeds, and a number of other breeds of horses and ponies, is shown in the table opposite, which gives the numbers of stallions-pedigree and non-pedigree-for which new licences were issued in 1939 and 1946 respectively. The great decrease in total numbers reflects the progress of farm mechanization. The figures were supplied by courtesy of the Government Departments.

HORSE BREEDING

Breeding animals should be chosen primarily for individual merit, and should preferably be of pure breed and of good pedigree. Continued and indiscriminate crossing should always be avoided as it gives a greatly increased chance of" misfits." In the breeding mare, quality, trueness to type, and roominess of body are more important than size (height). Freedom from any hereditary tendency to unsoundness is most important. The list of diseases that are held as disqualifying stallions from registration under the Horse Breeding Act includes cataract, roaring or whistling, bone spavin, ring-bone, side-bone, navicular disease, shivering, stringhalt, and defective genital organs.

Fillies may be mated for the first time either at two or. at three years old. The question whether it is advisable to breed at the earlier age must be answered according to circumstances according to the animal's size and condition, and whether it is desired that she should make the fullest growth, of which she is capable. Colts at two years old may serve a limited number of mares-generally not more than a dozen. Three-years-

Breed and Type	**England and wales**		**Scotland**	
	1939	**1954**	**1939**	**1953**
Heavy—				
Shire	1092	29	1	...
Clydesdal	215	5	459	20
Suffolk	255	12	2	...
Percheron	92	14	6	...
Cleveland bay	...	2	...	...
Total heavy horses	1699	62	468	20
Light-				
Arab	15	14	1	1
Thoroughbred	166	29	4	2
Hackney	...	24	...	12
Hunter	...	...	2	...
Other	28	...	...	...
Total light horses	209	67	7	15
Ponies and Cobs-				
Dales	13	1	1	...
Fell 7	1	...	1	
Highland	1	2	25	8
Polp and ridind	13	5	1	...
Shetland	5	2	10	9
Welsh pony	7	19	...	2
Welsh cob	41	1	...	...
Others	...	3	...	...
Total ponies and cobs	87	34	37	20
Others	...	...	34	1
Grand total	1995	163	546	56

Number in Northern Ireland (1953) were 24 Clydesdales, II Thoroughbreds, and 34 of other breeds including points.

olds should be restricted to about sixty and aged horses to about eighty. Much, however, depends on the individual characteristics of the stallion, on his constitution, and the effectiveness, of his services.

The period of gestation averages rather more than eleven calendar months and varies quite commonly as much as a fortnight ei-ther way. It is longest in mares that foal in winter or early spring and shortest in those that foal in summer. The first heat begins four to seven days after foaling and ends on the tenth to the thirteenth day after foaling. The average duration of the period of heat is seven days, but there is much variation. It is longest in early spring-and especially in a cold spring-and gets shorter in summer and autumn. The average time between successive heats is twenty-one days, but the best way to calculate the time of onset is to count sixteen days from the end of the previous heat. The foaling season may be at any time between January and September. With Thoroughbreds, where age is counted from the beginning of the year, and where two- and three-years-olds are very often trained and raced, it will obviously be advantageous to have the foals early, and most are dropped between January and March. For ordinary farm horses in this country the most generally convenient season is April to June, since this entails a minimum of housing and attention for the young animal. In countries where summer work is heavy and winter work very light, it is some-times arranged to have a proportion of the foals dropped in autumn.

In general, the fecundity of the mare is far below that of other domesticated animals. An occasional animal conceives regularly, year after year, at her first service; but it is far commoner for mares to "come back" to the stallion several times, and the proportion that remain barren after repeated. services throughout the breeding season may frequently reach 40 or 50 per cent. A service two to four days before the end of the restrous period is much more likely to be effective than one in the first day or two or on the last day of the heat. Indeed, if a mare is still on heat three days after service it is advisable to serve her again. The proportion of conceptions depends partly on the male. Some stallions will "settle" 80 per cent, of their mares, while others, under identical circumstances, will beget very few foals, and a small proportion of apparently normal animals are almost or quite sterile. To ensure good results the stallion must be carefully fed on a good mixed ration, should not be too fat at the commencement of the breeding season, and should have regular exercise during the idle part of the year. As regards the mares, those at grass are more likely to conceive to those kept in stables and on dry feeding. Mares that are on hard work and lean, and on the other hand, such as are over-fat, are much less likely to conceive than such as are in natural thriving condition Twins are rare and are not wanted, since they are generally undersized even when mature.

Ordinary farm mares are generally required to work regularly during the greater part of the year. It is only when the prospective value of the foal is very high in proportion to the cost of maintenance of the mare that breeding, in itself, is likely to produce a profit. Provided tot the mare is in the charge of a skilled and reasonably careful horseman, she may do all classes of farmwork during the first six or seven months of pregnancy. In the latter part she should not be put between shafts and should be excused from any exceptionally heavy jobs, but may continue to work in chains right upto the time of foaling. It is a serious mistake to rest the mare for the last few weeks, if this entails shutting her up in a box and leaving her without exercise; but if it is possible to turn her out to grass, she may go idle without taking harm The approach of parturition is indicated by a gradual slipping or relaxation of the muscles about the tail head and a swelling of the udder; generally some two days before foaling a waxy secretion will appear on the ends of the teats. As the time approaches, the mare should be placed at nights in a roomy box, and after the appearance of the wax should be kept under close observation. The period of labour is generally short-often less than an hour–and false presentations are comparatively rare. Retention of the cleansing is more dangerous than with other animals.

Some patience is frequently necessary in getting the foal to suck, but interference should be as long delayed as possible. After foaling, the mare should have laxative and nutritious foodsbran mashes, oatmeal drinks, and clover hay. After two or three days, if the weather be genial, mare and foal may begin to go to grass for a short period, which period is extended day by day until after two or three weeks they may be left to lie out at nights. In the case of foals bora early in the year, exercise should be regularly given until the weather conditions permit of the animals being turned out. When the foaling falls in summer excellent results are often obtained by turning the mare into a clean and sheltered paddock and allowing her to remain out altogether. Where foaling boxes are used, they must be thoroughly disinfected for each new case and must be kept clean.

After foaling, the mare should be left idle for at least a month, and may be so left with advantage until the time of weaning. When work is pressing she may be given light jobs, but should not be overheated nor kept apart from the foal for more ton two or three hours at a stretch. If, occasionally, the mother should be put to severe work or overheated, some of the milk should be drawn by hand before the foal is allowed to suck. The age for weaning may be about four months-or upto six months, if the mare is not required for work. The separation should be complete and final; the foal must be closely and securely confined out of hearing of the mare, and the latter should be put into regular work in order to dry off her milk. After three or four days' confinement, during which time it is usually fed on bran, oats, and hay, the foal may go back to pasture-preferably with a companion.

If weaning has occurred fairly early in the year, and if good pasture is available, no additional food will at first be necessary in order to keep the weaned foal in thriving condition. Backward animals, or such as have been weaned at an early age, will benefit greatly from a daily mash of bran and crushed oats soaked with separated milk. As the season advances and pastures get bare and less nutritious, the foal should get one feed per day, and later, during the winter, two. A suitable ration for midwinter might be: crushed oats, 4 lb.; bran, 2 lb.; and chaffed hay, 4 lb. fed either dry or slightly damp and divided into two meals. During periods of frost and snow a little long hay may be thrown out in the field, and many breeders give two or three swedes a day. The foal should be out whatever the weather conditions for some part of each day, but should-except in mild districts or in sheltered situations-be housed at night in a freely ventilated box. With the return of grass in spring, hand feeding may be discontinued, and in April or May the animal may begin to lie out.

Castration is generally performed at a year old or as soon thereafter as the weather is sufficiently mild. Very warm weather should be avoided. The operation involves more danger than with the other species of farm animals but is no less necessary, although some individual stallions, notably Percherons, work fairly well. No one but a skilled veterinary surgeon should be trusted to carry out the operation. Apart from this the only special attention necessary is to the feet. These should be kept well rasped down so that the frog rests on the ground, otherwise the feet will become long and narrow and an undue strain will be put on the pasterns. On ordinarily soft land, the feet will require to be dressed at intervals of four or six weeks.

Summer pasture for young horses should not be of the rich fattening type, since on such the animal is liable to get too heavy for its legs and thus lose activity and freedom of movement. Young horses grow and thrive best when they are given a wide range of moderate pasture and when they are running with other stock, preferably cattle. When kept by themselves on a small area, they graze very irregularly and are also much more liable to suffer from intestinal worms. Only Thoroughbreds are normally given grain (oats) while on summer pastures, and then only to hasten growth for early racing. During the second winter, if fair rough pasture be available, hand-feeding may be confined to a shorter period-say January to March and the ration may be lighter and less concentrated. Six pounds of grain with a similar amount of hay may ordinarily suffice, but the quantity must depend on the animal's condition and the climate. In no case should the animal be made fat, and it will usually be all the better, if it reaches the spring in distinctly lean condition. During the third summer ordinary pasture will again suffice.

The usual age for breaking the farm horse is two and a half years. Some Shires are put to work at two, and Clydesdales, in breeding districts or where the winter work is light, are often left till nearly three. In any case the whole process should not be left until the animal is of age to start work. As a young foal it should be made to understand the use of the halter and that there is no means of escape from it. This first lesson may involve a severe struggle. Thereafter, it should be gradually accustomed to being handled-to having its legs lifted, etc. When these processes have not been gone through at an early age the preliminary "gentling?' of the grown animal is much more difficult.

The first stage of the breaking proper is to put on harness and lead the animal about, in order that it may lose the sense of irritation and alarm at the contact of the tackle. Thereafter, it should be driven on long reins for an hour or so at a time, until it answers to the bit; two or three lessons will generally suffice. Care must betaken to avoid damage to the mouth, as repeated bleeding, even if slight, causes the "bars" (on which the bit rests) to become insensitive. These lessons may be combined with or followed by training in draft, which is best started by means of men pulling against the animal on long rope traces, the pressure on the shoulders being applied gradually while the animal is in motion. Finally, it may be harnessed along with a steady but active horse in the plough, with at first an additional leading rein and a man alongside. Work in shafts should be postponed for a few months. Judgment must be exercised in the matter of the time to be allowed for each stage. Some animals can be put almost straight into work while others may require a week or two of preliminary training. Fractious animals should be trotted round on the end of a rope-preferably on heavy going, such as fresh-ploughed land-before each lesson. This preliminary exercise should proceed until the animal is tired enough to be amenable to discipline, but not, except in extreme cases, to the point of exhaustion.

There is a good deal of variation in the methods by which the necessary horses are maintained on the farm. In the remoter country districts many farmers work their animals only until they are five or six years old, at which age they are sold for dray work in cities or to other farms where there is a large amount of road and cart work to be done. Provided that skilled horsemen are available for breaking and training, this scheme has the abvious advantage that the average value of the horses may be expected to appreciate during their period of farm work. Most horses continue to grow in size and strength until they are about seven years old, and if the whole working life is utilized a good animal, barring accidents, may be expected to remain fit for regular work until the age of from fourteen to eighteen years, after which it may till the post of "odd horse" for two or three years more. When all things are allowed for, the average working life will not greatly exceed ten years, say' from three to thirteen years old. The average life is influenced largely by the skill shown in feeding and general management, but. also by the amount of road work that tails to be done and, naturally, by the breed and individual character of the horses used.

FEEDING OF WORK HORSES

In Britain, oats and hay form the usual basis of the ration for work horses; and provided the oats be sound and not too new, and the hay be clean and hard, there are no other feeding stuffs that are so wholesome or so generally useful. Of the various hays, timothy, rye-grass, and good meadow hay are all very satisfactory. Clover is less good, since it is too laxative for animals on severe work. Dusty hay is objectionable in that it spoils the "wind," and mouldy or ill-got material should be avoided, if possible. For the winter months, when work is often irregular and the day is short, oat straw is often substituted for hay, and in such circumstances forms a useful fodder, though it is too low in nutritive value for animals on hard continuous work, in northern clay-land districts bean straw is used as a partial or even complete substitute for hay throughout the year. The taste for it is somewhat slowly acquired, but ultimately many horses come to prefer it to other fodders. It is less nutritious than hay, and the concentrated ration must be correspondingly heavier.

Of concentrates other than oats, maize is quite largely used, chiefly because it is often proportionately cheap: 3 Lb. give the same energy value as 4 Lb. of oats. If it is to form a large part of the ration, it must be introduced gradually, and it gives the best results when mixed with oats and dry bran in order to lighten it. Barley makes a good horse com but wheat is unsatisfactory, being liable to produce colics unless fed with care. Rye is worse. Beans are good upto a moderate proportion-say one-eighth of the grain ration. Crushed linseed, or linseed-cake, is an excellent conditioning food for animals tot are run down, but it is too laxative to be fed in quantity to animals on severe work; 1 or 2 Lb. of the former or 2 to 4 Lb. of the latter are common amounts to feed .daily. Molasses or one or other of the various molassed foods is sometimes added as an appetizer to the grain ration, and is useful for shy feeders. A bran mash in place of the usual evening feed of grain is commonly given on Saturday nights as a matter of routine, in order to prevent, by its mild laxative action, the possible ill effects of the following day's idleness. It is specially important when work during the preceding week has been continuous and hard.

On farms where summer work is not very heavy the work horses usually run at grass, and are given a feed of 5 or 6 Lb. of concentrate for each half-day that they work They must be stabled and "hardened up" before the regular autumn work commences.

The amount of the daily ration must be determined according to the size of the animal and the amount and severity of the work being done. A

Mi-sized, active, heavy, draft animal, on hard and continuous work, will require quite 20 lb. of oats-or its equivalent in other grain-per day, along with nearly the same weight of good hay. The following are the examples of rations for a farm horse of average size-say 1400-1500 lb. live weight:

Light or Intermittent Work	**Medium Work**	**Continuous Heavy Work**
12 lb. oats.	16 lb. oats.	10 lb. maize.
8 lb. hay.	16 lb. hay.	5 lb. oats.
10 lb. oat straw.	10 lb. swedes.	5 lb. bran.
20 lb. swedes.		14 lb. hay.

The actual expenditure of energy depends to a greater extent on the speed at which the work is performed than on the hours worked or the pull. Hence, in the north, where it is the custom to drive horses rather hard, rations tend to be heavier than in the south, in the Midlands and South of England, 12 lb. oats, mixed with 4 lb. of straw chaff and about 14 to 16 lb. of hay, may be taken as an average ration.

In periods of complete idleness the grain ration must be severely restricted, as many ailments arise from overfeeding at such times. In general, of course, the rations must be adjusted to meet the individual requirements of the animals, the object being to keep them in good, hard condition but not to make them fat.

As regards the preparation of the food, the grain should be bruised if possible, a proportion of the fodder should be chaffed, and grain and chaff fed together. This has the double object of preventing the animal from bolting the grain and of saving part of the work of mastication Some horsemen give the whole of the food in the form of a mixed chop, but the consensus of opinion is that some long fodder is advantageous. The old practice of boiling or steaming roots, grain, and chaff, and feeding the mixture in the form of a warm slop, had nothing to recommend it, and has been very generally abandoned. Feeding should be done not less than three times a day, and at least a full hour should be allowed both for the morning and for the mid-day meal. Water should be given freely four times a day-always before feeding, or, at least, not immediately after a meal. Exception should be made only in the case of an animal in very overheated and overtired condition; in such cases the horse's thirst should be quenched gradually with half a bucketful of lukewarm water at short intervals until he is satisfied.

Horses at work should be thoroughly groomed once a day. Legs should not be washed often, but should be allowed to dry and be then brushed out. In periods of idleness during winter an hour's exercise should be given at least every second day. Exercise is the more necessary, if the animals are suddenly thrown idle after a spell of hard work.

Shoeing should be done at intervals of six or seven weeks, or oftener in cases where the feet are thin or liable to break. New shoes are, of course, not necessary on all occasions. At the onset of winter, but before the coat is fully grown, farm horses should be clipped "trace high" -*i.e.* the coat is removed only from the chest, belly, and thighs, while the head, neck, back, quarters, and legs are left untouched. This plan allows the horse to keep reasonably cool at work and yet leaves the natural protection against rain on the one hand and mud on the other. The end of November is generally a suitable time to clip, and the operation is usually repeated in January. The stable should have ample head room, and should be well-ventilated and well lit. Stalls should be 6 ft. to 6 ft. 6 in. wide, and the length from manger to back wall should be about 18 ft. Overhead racks are undesirable because of the risk of hay seeds, etc., getting into the eyes.

Careful routine management must be accompanied by a close watch over individual animals, and early attention must be paid to common ailments, such as sore shoulders, "weeds" (lymphangitis), colics, colds, and lamenesses. Horses that are in lean and unthrifty condition should have their teeth examined and filed if necessary. Intestinal worms are another frequent cause of a failure to thrive. A supply of rock-salt in the manger will generally help in the maintenance of health.

ASSES AND MULES

Donkeys are used in this country for farmwork, but are generally to be found only on the smallest class of holding, for which they are rendered suitable by their small food requirements. Abroad, they are used as riding and pack animals. The larger breeds, which produce jacks suitable for mule-breeding, are the most valuable. The breeds that are held in the highest esteem for this purpose are the Catalonian (Spain) and the Poitou (France).

The former has unusually good symmetry and action and reaches a height of 14 to 15 hands; the latter is a breed of great weight and substance and large bone, and is generally from 15 to 16 hands high. The females are about 2 in. smaller in each case. Other well-known breeds are the Maltese, the Majorca, and the Andalusian. Mules are preferred to horses both for farm and city work in warm climates. They are very sure-footed, and hence, valuable as pack animals in mountain regions; they also possess great endurance and are not fastidious feeders. They are bred in great numbers in most Mediterranean countries and also in the warmer regions of both, North and South America. Big mules can be bred only out of big mares. The offspring of a horse and a female donkey (hinny) has its foetal growth restricted and remains undersized in after-life.

CHAPTER VII

PIGS

THE pig belongs to the non-ruminant section of the Artiodactyls (even-toed hoofed animals). The wild pig is a forest animal, feeding upon acorns, beech mast and other seeds, roots, worms, grubs, and occasional small mammals. It consumes relatively little in the way of grass or other herbage. The digestive system is closely similar to that of man, except that there is more active bacterial digestion (of cellulose) in the large intestine. The wild sow normally farrows only once a year and the average litter is about four.

All our domesticated swine are believed to have descended from two closely related species-Sus *scrofa,* indigenous to Europe and North Africa, and *Sun vittatus* of Eastern and South Eastern Asia. The latter is distinguished chiefly by its wider and shorter head.

The earliest known domesticated pig, *Sus scrofa palustris,* existed in Neolithic times, its remains being found along with those of the small short-horned type of cattle *(Bos brachyceros* or *longifrons)* and the small sheep of the same period. The domesticated pig of early historic times in Europe appears to have been closely related to the local wild type-long-legged with a long snout, heavy shoulders, short body, light hindquarters, and a coat of coarse bristles. During Saxon times large herds were maintained, the animals being herded, for a considerable part of the year, in the widely prevalent oak forests. More recently pig-keeping was largely associated with dairying on the one hand and with corn growing on the other. On butter-making farms the traditional foods were skim milk, wheat offals, and barley meal, and on cheese farms whey, wheat offals, and beans. On corn farms the chief foods, as would be expected, were tail corn and millers' offals, supplemented by beans or peas, In either case the breeding sows and young animals, wherever conditions permitted, had a run on pasture in spring and on stubbles during autumn. Still later, pigs came to be associated with commercial potato-growing.

During the latter part of the eighteenth and in the early nineteenth centuries, there was a considerable infusion of fresh blood by crossing with the so-called *Sus indicus,* introduced both directly from its home area in China and Siam, and indirectly through Mediterranean types, notably the Neapolitan. The Chinese differed from the native type in being smaller, shorter faced, and very markedly earlier maturing-*i.e.* with a much more pronounced tendency to fatten while young. The admixture resulted in an animal that was smaller and less active than the indigenous type and easier to fatten, but of less fecundity. Among modern breeds the Chinese influence is most obvious in the Berkshire and Middle White, least so in the Tamworth and Large White.

Most of our present-day breeds have been evolved, or at least have been moulded into their present forms, within the past seventy or eighty years. Previous to the formation (in 1884) of the National Pig Breeders' Association, there were no breed societies and no herd books. Up till that time breeding was conducted in a rather haphazard fashion, and although there were local types known as the Berkshire, Tamworth, Yorkshire, etc., their characteristics were very variable, and often widely different from those of the modern breeds that have inherited the names.

Pigs are, of course, valued almost solely as meat producers, but the meat may be utilized in different ways, and different types have been evolved to meet the varying requirements. Moreover, consumer demand and methods of utilization in Britain differ from those of most other countries. In the United States, for example, the hams, shoulders, and belly cuts are ordinarily cured, while the back and loin, after the removal of suplus back-fat are marketed in the fresh state and retailed mainly as pork chops. In Britain, before the war, there were four main market types viz.: (1) In the north, a relatively large and fairly fat pig, for bacon; (2) in the south, a lean and long pig of about 200 lb. live-weight (or 150 lb. carcass weight), for the Wiltshire bacon trade; (3) "cutters" with a much wider weight range than that demanded by the Wiltshire curer. The method of utilization was to cure the hams and sometimes the forehams and belly cuts, and to sell the rest of the carcass as fresh meat; (4) the London porker, of about 70 lb. carcass weight, which was retailed as pork joints. In all cases trimmings and much of the edible offal were normally converted into sausage, brawn, etc., while surplus fat was rendered into lard.

The weights mentioned above represent much less than mature weights i.e. they are for animals still in active growth. Good specimens of the larger breeds, at maturity but in breeding condition, range from 300 to 400 lb. live-weight in the case of sows, with boars about 100 lb. heavier. Exceptionally large animals, in very fat condition, reach weights of the order of 1000 lb.

Of all our farm animals, the pig has the greatest capacity to accumulate body fat, more especially in early life. Hence, early maturity., which is one of the chief objectives of the breeder of beef cattle and sheep, can, in the case of the bacon pig, be very easily carried to excess. It is, therefore, well that the selection of breeding stock, with the object of the right degree of early maturity, should be made when the animals have reached market weight e.g. in the case of bacon breeds, at about six or seven months old. An animal which attains the desired body proportions at, say, five months old on normal feeding, will be too short and blocky, and will tend to yield an over-fat carcass, when it reaches bacon weight.

The importance of the correct degree of early maturity is illustrated in Fig. 14. The typical Large White, slaughtered at 100 lb. live-weight, yields a carcass with too much bone and too little fat, but" when grown to 200 lb. gives a carcass of well balanced proportions. The typical Middle White gives a good blocky carcass at 100 lb. but, at the higher weight, is too short, too heavy in the fore-end, and altogether too fat for any purpose.

It is, of course, possible, by restricted feeding in the later stages, to delay maturity and, by full feeding on a specially nutritious diet to hasten fattening. But any major departure from normal feeding-more or less to appetite on a moderately concentrated diet-increases the food cost per pound of carcass weight.

The conformation desired in a pork pig, to be finished at four or five months of age, is comparable to that of a beef type of cattle or a mutton type of sheep. The highest-priced cuts are obtained from the back and loin (roasts and chops) and the ham (leg of pork). Hence, a wide, thickly fleshed back and a heavy ham are wanted. At the age indicated pigs are still in rapid growth and, if fed on a well-balanced diet, are unlikely to be over-fat. In the selection of breeding stock at porker weight a blocky conformation and thick fleshing should be the main criteria. Great length of body is incompatible with the degree of early maturity required.

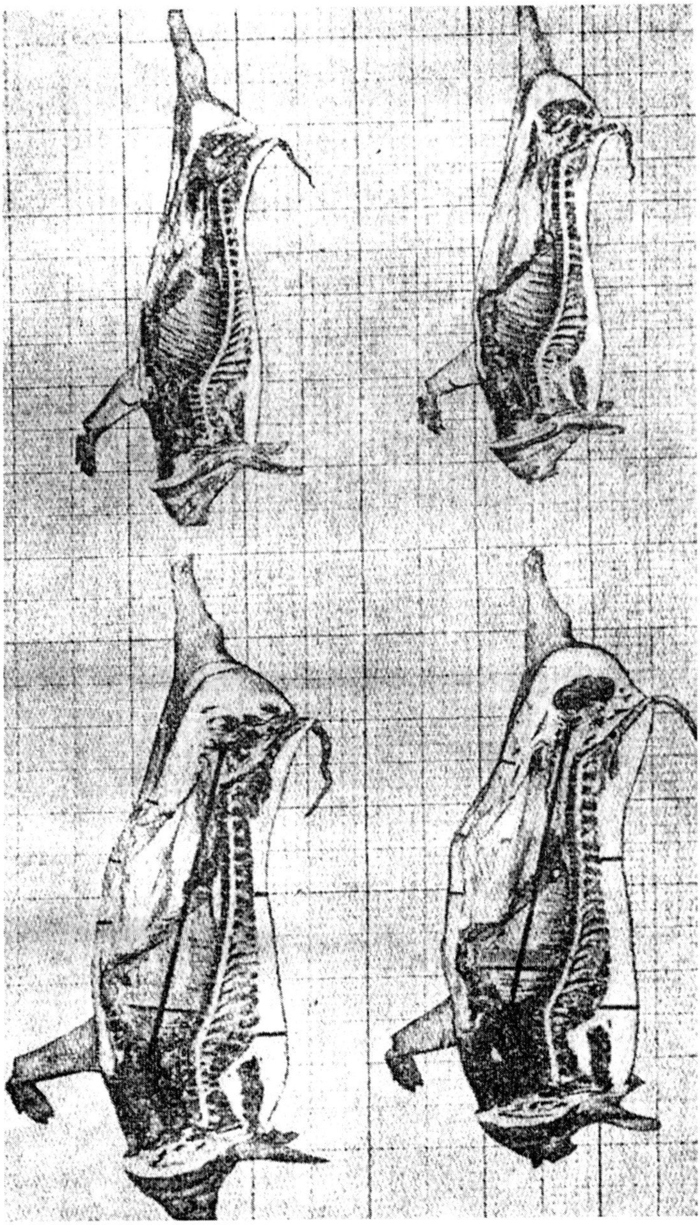

FIG. 14 : Carcasses of Middle White (right) and Large White (left) at 100 lb. live weight for (top) and at 200 lb. live weight for Bacon (below)

The bacon type is longer-bodied, rather higher on the leg, and more lightly fleshed than a pork animal of similar age. Since the side (back rashers and streaky bacon) and the ham provide the most highly priced cuts, great length, combined with a full, plump ham, are the main desiderata. The shoulder and jowl should be light. The side should be of uniform depth throughout and thick at the flank. In score-card systems of grading, length of body in relation to weight is emphasized, the other major points being a plump ham, a moderate thickness of back fat ½ in. or not exceeding 2 in.), a thick belly wall (streak), and a light head and shoulder. The firmness and whiteness of the fat are also taken into account, soft carcasses being penalized; but the quality of fat is determined by feeding (see p. 248) rather than by breeding.

The distinguishing characteristics of the bacon and pork types are to some extent fixed in the various breeds. The Large White and Tamworth are essentially bacon breeds. The average type of Berk-shire, being shorter, wider, and smaller of frame, is admirable for pork and useful for small bacon, but is unsuited to the production of full-sized bacon. Other breeds of intermediate type, with a cer-tain amount of selection and appropriate management, can be successfully used for either purpose. The more important British breeds are arranged by Hammond in the following groups, but in some there is a considerable variation as between strains :

Pork:

Middle White.

Berkshire.

Dual Purpose:

Essex.

Large Black.

Long White Lop-eared.

Wessex.

Welsh

Gloucestershire Old Spot.

Bacon:

Welsh.

Large White.

Tamworth.

Swedish Landrace.

PLATE 26

A. LARGE WHITE BOAR

B. MIDDLE WHITE SOW

A. BERKSHIRE GILT

B. LARGE BLACK SOW

PLATE 28

A. TAMWORTH BOAR

B. GLOUCESTER OLD SPOTS SOW

PLATE 29

A. LINCOLN CURLY COATED SOW

B. SWEDISH LANDRACE GILT

Cross-breds are frequently preferred for commercial purposes. Thus, the Berkshire x middle white cross is admirable for small pork, while a dual- purpose sow-e.g. the wessex-may be mated with the Middle white for pork or the large white for bacon.The arguments for and against cross-breeding are set out later (see p.239).

Apart from the special points desired for particular purposes, certain qualities are desirable under all circumstances. Robustness of constitution is one. Breeders of all classes of livestock tend to believe that "constitution" is correlated with certain points in conformation-for instance, a wide head or a capacious chest. In fact, there is no convincing evidence of any such correlation in the pig. Hardiness and thriftiness can be improved only by the continued weeding out of weakly animals or such as fail to attain a satisfactory rate of growth Another requirement is sufficient strength of limbs to carry the body and to enable a reasonable degree of activity, which is very important in the breeding sow. The legs should be reasonably stout and fairly straight and the pasterns short and nearly upright The temperament should be docile and quiet; nervous or savage or excitable animals are almost invariably unsatisfactory and, in particular, the temperament of the sow has much to do with the casualty rate among her young. Prolificacy is of obvious importance and, although there is a wide range of variation in the size of successive litters from the same sow, there is a considerable degree of heritability. Hence, breeding animals of both sexes should be chosen from among the progeny of prolific dams.

The growth rate of the sucking pig during the first month of its life (a point of great economic importance) depends partly on its inborn capacity for growth but much more largely on the level of milk production of its dam. Hence, sows whose litters bear evidence of poor nutrition should be discarded. Each pigling, within the first day or two after it has been farrowed, has its own teat; hence the maximum number that can be properly nursed is the same as the number of the dam's teats. In general, females should not be retained for breeding unless they have at least twelve well-formed and regularly spaced teats. Since teat number is inherited from both parents, the rudimentary teats of the boar should be examined with the same standard in mind. Gilts with ten teats, or even fewer, are not uncommon.

Most breed societies publish special registers of sows that have attained certain standards of performance in breeding and nursing-for example, an average for four successive litters, born within a period of twenty-two months, of eight piglings reared to weaning age.

Finally, colour is of some importance. On the one hand white breeds are somewhat liable to suffer from sun-scald when exposed to intense sunlight, though trouble of this kind is rare in Britain. On the other hand, black animals are liable to "seedy cut," *i.e.* the appearance of patches of dark colouration in the streaky rashers. The pigment can occur in the rudimentary mammary tissue of male pigs as well as in females.

Two procedures have been devised for the assessment of economic qualities in pigs and these should preferably be used in conjunction as guides to the selection of breeding stock The one is known as *pig recording,* records made being of the number of young born in each litter, the number reared to the normal weaning age (eight weeks), and the individual weights of the piglings at that age. It is a useful refinement to record weights at three weeks of age as well as at eight, because the growth rate during the first three weeks is a measure of the milking propensity of the sow; subsequent growth is largely influenced firstly by the quality of the food (especially on the amount of skim milk or other milk products included), and, secondly, on the skill exercised by the pigman in inducing the piglings to "take to the trough" at an early age. Several pig-recording societies were operating in this country in the thirties, and plans had been made, at the time of writing, for developments on a national scale.

Official recording commonly ceases at weaning age, but the pig-keeper will be well-advised to keep other records for his own guidance-of the age at slaughter of individual animals and of their carcass weights and carcass grades. The total food used, and the total live-weight increment of the herd, year by year, is a further useful measure of the overall efficiency of management. It is, of course, necessary to have a suitable weighbridge in order to arrive at the efficiency of food conversion even for the herd as a whole; but a weighing machine is a worth-while investment in any case since otherwise it is likely that individual animals, when sold, will fall outside the weight range to which the maximum price applies.

The other aid to improvement is *litter testing,* designed to assess the breeding value of individual boars. A reasonably accurate measure of a boar's merit can be obtained by selecting four of the heaviest weanlings (two gilts and two hogs) from each of three litters sired by him. The selected animals are moved to an official testing station, are kept under strictly controlled conditions, and are individually fed on a standard, balanced food mixture. Records are kept of food consumption, live-weights at slaughter, carcass weights, and scores for carcass quality. The average food cost (in terms of pounds of meal per pound carcass weight) and the average carcass quality of the twelve progeny, taken together, constitute a good measure of the sire's breeding value. In theory the same progeny test can be applied to sows, but the dam will be of rather advanced age before the test results for samples of three litters can become available; the results from any much smaller sample of progeny would be subject to a large margin of error. Litter testing has long been in use in Scandinavian countries and has proved an important aid to improvement both in conversion efficiency and in carcass quality. Litter-testing stations have recently been set up by the (British) National Pig Breeders' Association.

Until fairly recent times the ordinary basis of selection of breeding stock has been visual inspection, either by breeder or prospective buyer or show judge. Breed societies have recently, however, required members to provide returns of numbers born and reared in every litter from which any pigs are to be registered. It is, thus, possible to build up an extended pedigree of any animal, with all the evidence of the prolificacy of its female ancestors and of their ability to rear their young.

BRITISH BREEDS

There are, besides a few of small and purely local importance, eleven recognized British breeds, each with its own pedigree record. The Berkshire, Large White, Middle White, Large Black, Tamworth, Welsh, and Gloucester Old Spots are under the care of the National Pig Breeders' Association, while the Essex, Long White Lop-eared, and Lincolnshire Curly Coated have each their separate societies.

The Large White (or Large Yorkshire) (Plate 26, *A)* is derived from an early North Country type, large in size and either all white or white with black spots. Probably, it contained very little Chinese or Neapolitan blood. The evolution of the modern breed began, before 1850, in the industrial districts of the West Riding of Yorkshire, the pioneers being largely factory workers. In 1851, a weaver of Keighley, Joseph Tuley, exhibited at the Windsor Royal Show a pig of the improved type which attracted much notice.

The Large White is a pig of great size, being exceeded only by the Lincoln Curly Coat. Mature boars reach 7 cwt. in good breeding conditions and upto 10 cwt, when fully fattened. The head is of medium length and "dished," but the nose is not turned up. The ears are of medium length and incline forward but are rigid. The skin is fine and carries a moderately abundant coat of silky hair. The breed standard calls for pure white skin and coat, but occasional animals show small patches of dark skin and, more rarely, of black hair. The general conformation is that described for the bacon type-light in head, jowl, and shoulder, with great length of body, moderate depth of side, and plump but smooth hams. Broadly speaking, growth continues long and maturity is relatively late, but in this respect there is some variation as between strains. All strains can produce a first-class bacon carcass at the age of six to eight months, and some are quite suitable for the production of medium-weight porkers. All but the best types are somewhat inferior, in bacon conformation, to the Danish Landrace, which has been bred systematically for bacon quality for many years.

The Large White is, according to available records, the most prolific of all our breeds, the average number born per litter being over ten. But the sows are sometimes less docile than could be wished, so that losses of piglings, in the first few days of life, are higher than in certain other breeds. Milk production is good, and the young grow rapidly.

The Large White is widely distributed over Great Britain and Ireland, with some concentration in the north and east. It has been largely exported to other bacon-producing countries, including Canada and New Zealand, and has been used in the formation of many overseas breeds.

The Middle White (Plate 26, *B)* was formed by the blending of the blood of the Large White with that of the now extinct Small Yorkshire, which latter was a small, short-faced, fine-boned, and excessively fat animal, such characteristics reflecting the very high amount of Chinese blood that it carried. The Middle White was recognized by the Royal Agricultural Society in 1882, but the type was not fully fixed until twenty years later. Sanders Spencer, of Holy well, St. Ives (Hunts), played the leading part in developing the breed.

In weight the Middle White is about 20 per cent, less than the Large White and is probably the smallest of our breeds, though but little below the Berkshire and the Tamworth. The head is short, dished, and the snout turned up; the ears of moderate size and carried erect; the jowl is sometimes heavy and the general bold is rather short, wide, and compact; the breed is low-legged and the coat is thicker than in most other breeds. The Middle White, maturing very rapidly, formerly shared with the Berkshire the trade for small porkers, and the decline in its numbers is to be explained by the disappearance of this market. The modern tendency has been to breed for greater length of body, a lighter shoulder, and a leaner type of carcass. The sows are less prolific than those of the Large White, and the growth rate of the young, as would be expected, is less. Boars of the breed are useful for cross-breeding with sows of the larger and leaner breeds, and have proved very valuable overseas in the grading up of unimproved stock.

The Berkshire (Plate 27, *A)* originated, as its name implies, in the Thames Valley. Culley, writing in 1789, describes Berkshires as generally reddish brown with black spots. . . large ears hanging down over their eyes, short-legged, small-boned, and exceedingly inclined to make rapidly fat"; but other descriptions vary widely from this, and it seems that there must have been a great diversity of types. The modern breed is of a rather rusty black colour, with white "points" tail, face, and feet. It is, according to Smithfield weights, slightly heavier than the Middle White. It if; smaller-framed than the Tamworth but about as heavy. The head is rather short but the snout is not turned up. The ears are rigid but incline forward. Body conformation is similar to that of the Middle White, but the hind quarter is typically more level. The Berkshire is less prolific than the larger and longer breeds and the young sometimes make slow growth in early life.

The outstanding merit of the Berkshire is the quality of its meat at the small-porker stage of development. At one time pure bred Berkshires were supreme in the Smithfield pork-carcass competitions, but the breed has suffered, like the Middle White, by the loss of its special market. Abroad, the Berkshire is widely popular, its dark colour being an advantage in sunny climates. The types developed in the United States, Canada, and Australia are generally longer-bodied and leaner than the original.

The Large Black (Plate 27, *B)* is an old breed which came into prominence only after the founding of its Breed Society in 1899. The foundation stock came partly from Cornwall and Devon and partly from Suffolk and Essex, but the original differences in type have long ago been eliminated. The breed is widely distributed in Britain and has been largely exported, especially to warmer countries.

Animals of the breed reach weights only slightly less than Large Whites of comparable age. The head is of moderate length, the ears long and lopping, the tips reaching to the nose. The hair is jet black, fine, and generally rather abundant, but the skin is not deeply pigmented. The general build is intermediate between that of the Large White and Tamworth on the one hand and the Berkshire and Middle White on the other; in other words, the conformation is that of a dual-purpose animal. The shoulder is notably light in relation to the ham, but the flesh is often less smooth than that of the Large White. The sows are remarkably docile and seem to be able to build up bodily reserves, during pregnancy, with very little food other than good grazing. The average size of litter is about nine as compared with ten for the Large White, but the casualty rate is substantially less. The Large Black sow, crossed with the Large White boar, gives a product with much of the bacon quality of the sire and a mainly white colouration, and which has the advantage of the good ''mothering" ability of the dam.

The Tamworth (Plate 28, *A)* has its home in the Midlands, particularly the district about Birmingham. It is of comparatively small weight, long-bodied, and relatively narrow in build, with a very long straight snout and red (chestnut) colour. The breed was at one time reduced to small numbers, and prolificacy (probably through inbreeding) declined seriously, but importations from Canada have helped to restore fertility and vigour. The Tamworth is slow-maturing, but its long smooth side, neat shoulder, and firm flesh makes it very suitable for the production of high-quality bacon.

The Wessex Saddleback (Plate *31, A)* and the Essex (Plate 31, *B)* are of similar type and colour-black with a white "saddle" over the-shoulders and forelegs. The Essex breed-standard calls for white hind feet and a white tip to the tail, whereas in the Wessex the hind feet and tail should be black. Both breeds are liable to produce a proportion of all-black individuals, the probable explanation being that the wide-belted and all-black types breed true, while the animals with narrow belts are often heterozygous i.e. they throw wide-belted, all black, and narrow-belted progeny.

The ears are of medium size, with a forward pitch, but rigid. They have more forward inclination in the Wessex. Both breeds are hardy and the sows milk well. The Wessex is slightly the larger.

The Saddlebacks are, after the Large White, the most numerously represented breeds, and the cross between the Large White boar and the Saddleback sow is very popular with commercial breeders. The statistical evidence on prolificacy is to the effect that, although the Wessex has a lower average number of pigs born per litter than the Large White, the average number weaned is slightly greater.

The Gloucester Old Spots (Plate 33, *B)* is an old local type which was little known until the establishment of a herd book in 1914. In conformation it approaches the Large White. The pig is hardy, prolific, and a good grazer.

The Lincoln Curly Coat (Plate 29, *A)* is larger than the Large White more heavily fleshed, and rather coarser in bone. The coat is abundant and curly. **The Swedish Landrace** (Plate 29, *B)* is of extreme bacon type, very long-bodied and light in shoulder. The face is medium long and the ears incline forward. **The Long White Lopeared** (Plate 30, *A)* is, apart from colour, closely similar to the Large Black. Its home district is Devon and Cornwall. **The Welsh** (Plate 30, *B)* is similar, but rather longer and leaner in build. It is gaining in popularity by reason of its hardiness and good carcass qualities.

There is an almost complete unanimity that the number of British breeds is unnecessarily large. The needs of the industry could probably be met by half the present number-two bacon breeds, perhaps two dual-purpose types selected for prolificacy and mothering ability, and an early-maturing pork breed.

All boars are required to be licensed by the Agricultural Departments. The following table, giving the number of new licences issued in the last year for which figures are available, indicates the relative importance of the various breeds :

Breed	**England and Wales, 1954**	**Scotland, 1953**	**Northern Ireland, 1953**
Brekshire	101	...	...
Large White Yorkshire	19,050	1890	960
Middle White	125	...	...
Tamworth	102	8	...
Wessex Saddleback	1,371	210	...
Cumberland	5	...	4
Essex	1,031	34	...
Gloucester Old Spots	58	...	...
Large Black	620	91	...
Long White Lop-eared	55	...	...
Welsh	985	...	...
Landrace	251	...	...
Yorks Blue and White	6	...	...
Oxford Sandy and Black	2	...	...
Other Breeds	14	3	...
Total	23,777	2236	964

BREEDING

Well-grown and liberally fed gilts begin to come on heat at the fifth or sixth month of age, and from this time until they are large enough for mating they should be kept apart from boars.. The minimum age for breeding, for good results, is about eight or nine months in the case of the female. But the actual criterion should be size, and a body weight of about 175 lb. in store condition may be taken as a reasonable standard. Earlier mating often results in an unsatisfactory first litter and may cause permanent stunting of the sow's growth. Some breeders prefer to delay mating to an age often or even twelve months. Young boars (brawns) may begin to serve at seven or eight months of age.

The female comes on heat, except when pregnant or nursing, at intervals of twenty to twenty-one days throughout the year, so that it is possible to arrange for litters to fall at any season. Cestrus is rare while a sow is in milk, but occurs very soon after her litter has been weaned, very commonly on the third or fourth day. The period of heat lasts forty to sixty-five hours, and the onset is heralded, some two days in advance, by a swelling arid reddening of the vulva. Ovulation occurs towards the end of the heat period, so that sperm from a service soon after the onset may have lost its vitality before the eggs are shed. The ordinary rule is to mate during the second day of Cestrus and, in case the sow is still on heat the following day, to mate a second time. It should also be borne in mind that the first service from a young male, or from one that has been rested for some time, may be ineffective because of the inactivity of the sperm. In this case a second mating should follow a few hours after the first. Except under these exceptional circumstances there is nothing to be gained by mating twice during the same Cestrous period. Sows are more likely to conceive at the first heat after weaning than at subsequent heats.

Since the gestation period is sixteen weeks (varying only a day or two one way or the other), and since litters are normally weaned at eight weeks of age, the breeding cycle is completed in just under six months. Some sows, in fact, farrow at half-yearly intervals throughout their breeding lives, but the average interval between litters is between six and seven months.

It is sometimes argued that eight weeks is too early an age for weaning, but it must be recognized that the period of lactation of the sow is very short. In most cases the peak milk yield is reached about three weeks from farrowing, and the milk flow has commonly dropped to a low level by the eighth week. A surer method of preventing a set-back at weaning than that of extending the suckling period, is to induce the young to begin eating at the earliest possible age, and to use specially nutritious food during the suckling period and for a few weeks after weaning.

In large commercial herds, provided with adequate housing, the usual aim is to have the farrowings evenly distributed throughout the year and so to achieve a regular flow of marketable pigs. Any other system makes for periodic congestion of housing and to fluctuating demands for labour. But if the available buildings are poorly insulated, damp, or otherwise unsatisfactory, farrowing in the dead of winter is best avoided.

The useful life of a sow may ordinarily extend to four or five years or, say, six to eight litters. But since there is always a wide choice of gilts for replacement, sows tot have proved unsatisfactory should be discarded after one or two farrowings. Specially valuable pure-bred sows may be retained till an age of eight or even ten years, though they generally become progressively less prolific after their fifth year. Individual litters vary from one or two to more than twenty, and the average reaches about ten in the more prolific breeds. With first-class management numbers reared can reach an average of nine per litter or eighteen per year, but a figure of seven per litter, with an average interval of seven months between successive farrowings (giving an output of twelve progeny per sow per year), can be taken as satisfactory under the conditions of commercial farming. Gilts' litters tend to be one or two less in numbers than those of sows in their prime, but something depends on the age at first mating.

Very large litters are not desirable. More than twelve should not be left on a sow since, even if she has fourteen functional teats, the hindmost pair usually give very little milk. Many breeders limit numbers to ten for sows and eight for gilts, but with good stock and good management twelve and ten respectively may be permissible.

Surplus piglings, if healthy and vigorous, may be easily transferred to another newly farrowed sow, with a small litter of her own, if such happens to be available. A sow will commonly raise no objection to foster-piglings, even if these are of a different colour from her own. If no foster-mother is available the surplus pigs (being, of course, the poorest of the litter) must be killed; action, however, should be delayed until about the third day after farrowing, since most casualties occur during the first two or three days.

As regards systems of breeding, the simplest is, of course, to select a particular pure breed and to aim at progressive improvement by selection, with such aids as pig recording and litter testing or, at the least, by turning to account all the data about prolificacy, efficiency of food conversion, and carcass quality that can be got together. This is the system that has been applied with marked success in Denmark.

Many commercial producers in this country prefer a system of crossing. Where the process is confined to first crossing, the plan is that the sows should be of a breed combining prolificacy, milking ability, and "mothering qualities" with reasonably good conformation-e.g. the Wessex-and to cross with a boar likely to be prepotent for growth rate and carcass quality normally the Large White.

Apart from the advantage of combining the desirable qualities of the two parents there is the argument that first-cross animals are more vigorous and faster growing than pure-breds i.e. that crossing, as such, results in hybrid vigour. The experimental evidence in support of this contention is not very conclusive, a fact tot may be explained on the assumption that hybrid vigour depends on the heterozygous condition of certain particular pairs of genes, and that it will be largely a matter of chance whether across-bred animal will be more heterozygous, with respect to the factors concerned, than a pure-bred.

Sometimes crossing is carried a stage further with the object of taking advantage of hybrid vigour in the sow. Thus, one might cross the Large White with the Wessex, and mate the cross-bred females back to the Large White. Going still further, some American commercial producers follow a system of rotational crowing, using three breeds. Breed A is mated to breed B, the cross-bred sows to a boar of breed C, the produce of the last to a boar of breed A, and so on.

The obvious objection to all systems of cross-breeding is that either two pure breeds must be maintained or, alternatively, that the breeder must rely on others for a continuing supply of purebred gilts as well as of boars. In the latter case he cannot himself follow any system of progressive improvement.

Where the decision is in favour of a single pure breed, there remains the question of the degree of inbreeding that should be adopted; the matter is discussed in Chapter I of this section.

Still another system, also discussed in Chapter 1, aims at the production of highly inbred families, in each of which the least desirable genes have been bred out, and then crossing up two or more of these families for the production of commercial stock *i.e.* the same technique as is used for the production of hybrid maize. The process has been carried near to completion in the United States, and the production of inbred families is proceeding under the Animal Breeding Research Organization at Edinburgh.

SOME PRINCIPLES OF HOUSING

There is a great variety of housing systems for pigs. The choice depends on a balance of considerations, having regard to the local climate, the soil type, the size of the enterprise, and the available supply of capital and labour. The major objectives should be: (1) The provision of environmental conditions that will conduce to full health and vitality-warmth, comfort, and opportunity for the necessary minimum of exercise;

(2) the control of parasitic infestation and contagious disease; (3) the prevention of nutritional disorders-especially piglet amemia; (4) reasonable capital cost; and (5) economy of labour.

No system so far devised can be said to be the best under all circumstances and in all respects, and decisions must be reached. On a balance of considerations :

1. The pig, being a forest animal in nature, and thus having opportunity to seek shade in hot weather and shelter in cold, has a narrow range of tolerance in regard to environmental temperature. It should be remembered, too, that high air humidity intensifies the adverse effects both of heat and cold. Animals in show condition are subject to "heat stroke" and young pigs are very susceptible to cold. The commonest fault in farrowing sheds is that they are too cold in winter. A condition that has, in the past, been confused with piglet ancemia has been closely associated with chilling, due either to too Iowan air temperature (sometimes combined with high humidity) or to poorly insulated flooring. The optimum environmental temperature for a baby piglet, at rest, is probably about 80°F, a level that can be maintained inside a well-strawed nest in a building with an air temperature of about 65° F and with a well-insulated floor. Where the air temperature is liable to fall below 65° some form of artificial heating for the nest, should be provided. One possibility is floor heating by means of resistance wires; another is an electric heat-lamp suspended about 2½ ft. above the nest; or, failing a supply of electric power, an oil lamp. Space heating, by stoves or hot-water pipes, is unsatisfactory, because (apart from its high cost) high temperature throughout the building results in a high rate of ammonia production from urine.

All this may be thought to be at variance with the experience that piglets generally do well, even in winter, in the common type of simple wooden hut, of triangular section, with free access to the open; but it has been shown that the temperature maintained at the back of a well-bedded hut is generally much higher than that prevailing in a permanent farrowing shed with the usual amount of airspace and the common amount of ventilation.

Casualties in cold buildings are due either directly to chilling or indirectly through the tendency of the young piglets, in the absence of any other source of warmth, to crowd round the sow and thus, to risk being overlaid.

Where existing buildings have poorly insulated floors, a sleeping platform of wood or other non-conducting material may be provided. If the air-space is excessive a ceiling may be added at a convenient height.

A useful check on the adequacy of the temperature in the farrowing quarters is the seasonal variation in piglet casualty rates and in average weaning weights. Thus, if, in summer, the piglet death-rate is 5 per cent, and the average weaning weight 35 lb., and if the corresponding winter figures are 15 per cent, and 26 lb. a not uncommon relationship-the likely explanation is that the farrowing quarters are too cold.

Cold tolerance, of course, increases with body size but also depends on bodily condition, since the layer of back-fat acts as a blanket. Warmth continues to be of importance in piglets at the post-weaning stage and also in the case of sows in very reduced condition following lactation. Given a dry bed, well-grown stores, and mature animals, in good condition and run in sizable groups, require no special consideration.

As regards exercise, breeding sows during pregnancy, breeding boars, and gilts intended for breeding require a fair amount of freedom and should either be kept in yards or have an outdoor run. Stores and fatteners will generally get all that they need in sties.

Trough room is important both in order to ensure that all pigs in a group have equal opportunity to feed but also because overcrowding at meal-times may lead to fighting. One foot per head is the common allowance for pigs upto bacon size, but the optimum is probably rather more.

A very important measure is the provision of a " creep" in the farrowing pen, so designed as to keep out the sow and admit her young. Food, preferably in the form of pellets or small " nuts," should be offered in the creep from the age of three weeks.

Considerations of general comfort are important in relation to the choice between the indoor and outdoor systems. On free draining light land, a good sward can with ease be maintained in pig paddocks, and young pigs benefit from outdoor exercise. But with poorly drained soil in a wet climate the pen soon becomes a quagmire and so much mud is carried into the huts that it is impossible, in practice, to keep the interior dry.

2. As regards parasites, the most prevalent types-the lung worm, the round worm of the intestine, and the stomach wormare very liable to be picked up from contaminated ground, and unless the animals are frequently treated, fixed paddocks rapidly become "pig sick" Moreover, a sward stocked entirely with pigs deteriorates rapidly so that there is little grazing. Moreover, the manure is then largely wasted. Hence, permanent runs have to be periodically rested from pigs-i.e. must be stocked with some other type of animal.

The system greatly to be preferred is one in which the whole installation is moved at intervals of a few months to a fresh site with a clean and well-established sward.

There are three possible modifications of the hut-and-run system The one employs a fold unit-i.e. a hut with an attached unroofed pen, mounted on wheels or skids, which can be moved to clean ground at intervals of a few days or even daily. Under another system the sow is tethered by means of suitable harness and chain to a peg which is driven into the ground a few yards from the entrance to the hut, the young pigs being allowed free range over the pasture. A third is to confine the sow, by means of an electric fence, to a small area surrounding her hut, the height of the wire being, of course, sufficient to allow the piglets to range outside.

The wide dispersal of pigs under the outdoor system is a favourable factor in relation to the spread of certain contagious diseases. This is not true of swine erysipelas, which is caused by an organism that can live in the soil. In any case this trouble, as well as swine fever, can be controlled by immunization. Virus pneumonia (pig influenza), however, since the infection can be air-borne, is much more prevalent under the indoor systemat least where farrowing is carried out in a building that accommodates a number of litters.

3. The commonest nutritional disorder is piglet anremia, associated with a low iron content in the blood. The danger arises from the fact that iron is virtually absent from sow's milk, so that the piglet, though born with a certain reserve, soon becomes dependent on some other source. Green herbage and soil are the natural sources, so that piglets with access to the land are never affected. Moreover, if the pregnant sow spends part of her time out of doors, the piglings, at birth, will have considerable reserves. Under the indoor system the trouble can often be prevented firstly by feeding

greenstuff to the sow during pregnancy, and secondly by placing a fresh turf, or a spadeful of soil, in the farrowing pen daily. In other cases it is necessary to resort to dosage of the piglings, in their second week, with iron salts.

4. The prime cost of housing is, of course, greater in the case of a permanent building than that of movable huts, together with pig fencing for the surrounding paddocks. On the other hand, the rate of depreciation on the latter is much the higher.
5. Labour costs are reduced by concentrating pigs in large groups, with well-planned arrangements for feeding and the removal of dung. The feeding of scattered groups involves a large expenditure of time. Moreover, it is very difficult to achieve the proper degree of supervision, particularly of farrowing sows, if these are in dispersed individual huts over a large area.

A few miscellaneous points may be added.

Some breeders are accustomed to use farrowing crates in order to restrict the movements of the sow and prevent the young being trodden or overlaid, but it seems that the desired result can generally be achieved by providing a warm nest to which the sow has no access. Opinions differ as to the value of farrowing rails, which have the object of saving piglings from being caught, when the sow lies down against the wall of her sty.

While strong pig-netting is necessary to confine a boar, breeding gilts and pregnant sows, running in groups, can be controlled by an electric fence.

The usual size of farrowing huts is 8 by 10 ft., while indoor farrowing sties should ordinarily be 10 by 10 ft. Where a number of farrowing or feeding pens are grouped together in the same building, each, as a precaution against the spread of disease, should have independent drainage.

The housing system need not be either a purely indoor or a purely outdoor one. One popular arrangement is to have a block of farrowing pens in a permanent building with small runs to which the animals are given access for a short period every fine day; to have a yard or two, opening on to a large paddock, for breeding gilts and in-pig sows; to use an all-indoor fattening shed with a series of pens of varying size, so that groups of weanlings, as they grow, can be moved from the smaller to the larger, while backward pigs, when their sty-mates have gone for slaughter, can be placed in the small pens. Re-grouping should be avoided, so far as; possible, since this leads to fighting. Sows in very reduced condition at weaning should be kept by themselves until they have fully recovered strength and activity. Such needs should be foreseen in planning.

FEEDING

It must be borne in mind that the digestive organs of the pig are of a very different type from those of tile purely herbivorous animals. The capacity of the alimentary canal is, in proportion to the size of the animal, about half what is found in the ox, sheep, or horse. There is no provision, such as is made in the paunch of the ruminant or the crecum of the horse, for holding a large quantity of bulky, fibrous food while it undergoes bacterial digestion. Despite the smallness of its digestive tract the pig, on full diet, eats more, in relation to its size (and in terms of dry matter), than other farm mammals. This implies that the process of digestion is much more rapid. Bacterial digestion does occur in the large intestine, but this is capable of dealing with only a small amount of cellulose. Hence, the proportion of fibre in the diet must be kept low.

While cellulose digestion is much less efficient than in the ruminant, the more readily digestible materials-starch, sugar, and protein-just because they are not subjected to bacterial action, are more efficiently utilized by the pig.

The pig's capacity for growth and fattening is remarkably high since its high energy intake is coupled with a maintenance requirement that is not substantially greater than those of the other farm animals. A steer of 9 cwt, on frill feed, consuming 25 lb. of dry matter, with some 13 lb. of starch equivalent, may make a daily live-weight increase of about 2 lb., whereas a pig of 200 lb. weight, on a ration piovidiug 6 lb. dry matter and about 4 ½ lb. starch equivalent, may be expected to gain about 1½ lb. per day. Moreover, the proportion of carcass weight to live weight is substantially higher in the pig. Such comparisons must not indeed be pressed too far, since the steer can utilize straw, hay, pasture grass, and other cheap, bulky material in relatively large quantities.

In general, if the pig's capacity as a converter of feed into meat is to be fully exploited, its rations must provide a large quota of energy in small bulk. It is only in the case of the pregnant sow or the full-grown boar that the concentrate can be economically cut down by any large

PLATE 30

A. LONG WHITE LOP-EARED SOW

B. WELSH GILT

A. WESSEX 'SADDLEBACK' SOW

B. ESSEX SOW

PLATE 32

A. SECTION OF GLOUCESTER INCUBATOR

B. "SONDIX" BROODER

A. MOVABLE GENERAL-PURPOSE POULTRY HOUSE

B. .FOLDING UNIT

amount and replaced by grass or green food. Considerable use can, however, be made of roots, especially potatoes and fodder beet, since these contain only small amounts of fibre.

The fact that pig? require a highly concentrated diet has led, at times, to serious mistakes in feeding. The traditional foods barley meal and wheat middlings-are poor in regard to both the quantity and quality of their proteins and have a very unsuitable ash balance-a deficiency in calcium, an excess of phosphate, and a shortage of common salt. The traditional supplement was skim milk (or, alternatively, whey with beans or peas), and when this was used in adequate amounts good results were obtained.

Young pigs-upto the age of tour or five months-make maximum growth only if they have a proportion of animal protein in their diet; the usual sources are white-fish meal, meat meal, meat-and-bone meal, and separated milk either in liquid form or dried. Skim milk appears to have some special merit in promoting health and growth. After the age indicated, quite satisfactory growth and development can be obtained on rations containing vegetable proteins only. The oil content of the ration should be kept fairly low, especially in the later stages of fattening, in order to avoid the carcass fault of soft fat. Unsaturated oils are particularly objectionable, so that herring meal should be used in small amounts only. White-fish meal and meat-and-bone meal contain substantial amounts of calcium, so that, when used in considerable amounts with a moderate amount of clover or other greenstuff, the addition of minerals may be unnecessary. Most proprietary pig meals contain added minerals, often in great variety; but it seems that, in at least the vast majority of circumstances, it is sufficient to add about 1½ per cent, of ground chalk and ½ per cent, of common salt to the meal mixture.

Iron deficiency, as mentioned earlier, is the cause of nutritional piglet amemia, which may develop in cases where the piglets have no access to soil before they begin to take solid food-i.e. during the second and third weeks of age. The symptoms are commonly most severe at about the tenth day. It can be prevented by administering daily, during the second week of life, a small dose of ferrous sulphate. A convenient method is to dissolve 3 oz. of ferrous sulphate in half a gallon of water and add half a gallon of molasses, and to dose at the rate of a teaspoonful per day.

As regards vitamins the only likely deficiencies will be of A and D_3, the former only where insufficient greenstuff is fed, and the latter only in animals that get little or no direct sunlight or " sky-shine." In case of doubt on either score, about 1 per cent, of cod-liver oil may be added to the mash once or twice a week.

The total energy value of the food, in relation to its dry-matter content, is of obvious importance. For growing and fattening animals and for lactating sows the starch equivalent should be about 75 per cent, of the dry matter, or about 65 per cent on an air-dry basis. Sucking pigs and those newly weaned should have a rather more concentrated diet-say, a mixed meal with a starch equivalent of perhaps 68. For pregnant sows and growing gilts a figure of 55 will suffice, this being ordinarily arrived at by feeding a limited amount of meal along with grazing, or with greenstuff such as kale, cabbage, clover, or roots. With good clovery pasture only a small allowance of concentrate is necessary.

The protein balance should be varied according to the age, etc., of the animal. The young pig (up to the age of about four months) and the nursing sow have high requirements-a ratio of protein equivalent to starch equivalent of about 1: 5, or, in other words, about 13 per cent, protein equivalent in the mixture. Thereafter the protein equivalent may be reduced to about 11 1/2 per cent, and finally, in the later stages of the bacon pig's life, to about 10 per cent.

The following are the examples of mixtures suitable for the three groups in a breeding and feeding herd :

	I Nursing Sows and Weaners	II Growing Pigs 3 to 5 months	III Over 5 months of age
White-fish meal	7	4	...
Bean meal	7	9	13
Barley meal	25	30	35
Flaked maize	25	25	20
Weatings (Middlings)	35	30	30
Mineral?	1	2	2
Starch equivalent	68	67	67
Protein equivalent	13	11.5	10

A variety of ready-mixed foods are, of course, available on the market, either as meals, feed nuts, or pellets; and where the requirements for a particular group of pigs are small, there is a strong argument, on grounds of convenience, for their use. Where home-grown and purchased foods are used together, it is obvious that the mixture should be balanced. Too often a purchased compound is supplemented by tail-corn or potatoes, when, of course, the protein balance is seriously altered.

The "creep" feeding of sucking pigs is discussed later (see page 253). A common variation of all-meal feeding is to use a mixture like No. II above up till the point where consumption reaches 3 or 3 ½ lb. per day and thereafter to keep the meal ration constant, adding boiled potatoes, sliced fodder-beet, or sterilized household swill to appetite. Since all these materials haye a low protein content, the overall nutritive ratio of the combined ration falls progressively with the proportion of meal to other materials. This scheme is known as the Lehmann system, after the German worker who devised it. As already said, pigs should have a small daily allowance of green food, either in the form of pasturage or kale, green clover, etc. In the absence of a supply a small amount of dried grass may be incorporated in the meal, or be fed in the form of grass nuts. Under many circumstances, it appears that the incorporation of a small amount of antibiotic material-usually aureomycin or procaine penicillin-increases growth rate and the efficiency of food conversion. Typical results are 10 per cent, faster growth and 5 per cent, improvement in conversion. The main explanation appears to be that the bacterial flora of the digestive tract are controlled and minor digestive upsets are, thus, prevented, but the precise mode of action was not understood at the time of writing.

With regard to the preparation of the food, all hard grains must be either ground or soaked, otherwise they may escape digestion. Experiments in the feeding of dry meals (as against a thick slop) have shown no constant or important differences, but there is more risk of waste by dry feeding. Except for the farrowing sow, the slop should be of thick consistency, especially in cold weather. If dry meal or a thick slop is used, water, if not provided in drinking bowls, should be offered twice daily; but with a mixture of one part meal to two parts of water (by weight) it is generally found that pigs will drink only during the periods of hot weather. Cooking should not be done except in the case of potatoes (whose starch grains in the raw state tend to escape digestion) or in that of materials of doubtful soundness. Household swill must, by law, be cooked as a precaution against foot-and-mouth disease.

Where pigs are intended to be fed to appetite, the dry meal may be offered in self-feeders, of which several types are procurable, water being provided from troughs or drinking bowls. In general, growing pigs are frequently given all that they will eat, and there is little or nothing to be gained by restriction. On the other hand, breeding gilts and pregnant sows will get much too fat unless the concentrate is restricted; moreover, it is generally economical to compel them to consume a large proportion of grass, greenstuff, roots, etc. Again, all but the more extreme types of bacon pigs tend to get too fat to command the best price if, in the last month or two, they are fed *ad lib.*

During the last few weeks of the bacon pig's life the effect of feeding on the quality of the carcass should receive consideration. On the one hand, excessively watery foods-roots, whey, or distillery paste-should be fed only in moderate amounts. On the other hand, materials that contain any considerable amount of unsaturated oil should be avoided as likely to produce soft carcass fat. Examples are maize germ meal and, to a less extent, maize itself rice meal, and linseed cake. Fish meals derived from herring, mackerel, and other "non-white" fish impart a taint to the flesh. The best finishing materials are barley, potatoes, and beans.

Food consumption is ordinarily reckoned in terms of dry meal, 4 lb. of cooked potatoes, or about 5 to 6 lb. of fodder beet, or 8 lb. of swedes being reckoned as I lb. of meal equivalent. On this basis a pig of 50 lb. live weight will consume about 2 $^1/_2$ lb. of meal (5 per cent, of its weight); one of 100 lb. about 4 ½ lb. ; 150 lb., 5 ¼ lb.; 200 lb., 6 ¾ lb.; 250 lb., 7 ½ lb. (3 per cent of the body weight). A pound of meal per month of age is a useful approximation. A common practice in production for Wiltshire bacon, at 200 to 230 lb. live-weight, is to feed to appetite until consumption reaches 6 lb. per day, and thereafter to restrict the ration to this amount.

GENERAL MANAGEMENT

Pregnant sows should have ample opportunity for exercise, preferably in the open air and, in summer, on pasture. Many unsatisfactory litters, and many casualties among the young, result from sows getting fat and lazy. In summer a rough but rainproof shelter, with access to pasturage, can provide for a group of upto twenty. The same shed, with perhaps some additional provision against wind or driving rain, can be used in winter if the land is reasonably dry: otherwise a partially open yard is preferable.

Sows that finish lactation in a seriously reduced condition should have a short period of good feeding before they join the group, otherwise they may be bullied and prevented from getting their full share of food. The requirements of a pregnant sow, for maintenance and the rebuilding of body reserves, are met by about 6 lb. of meal equivalent. Good ordinary pasturage, or an abundant supply of good greens tuff or roots, will meet about half of these requirements, so that the actual allowance of concentrates may be about 3 lb. per day. Indeed, on very good clovery pasture satisfactory condition may be restored with 2 lb. of concentrate or, occasionally, less. For sows on pasture, feed-nuts are more convenient and less wasteful than dry meal or slop; these will be fully consumed if scattered thinly over the ground, and all the animals will have the opportunity of getting fair shares. On good clovery pasture or on lucerne a purely carbohydrate food will suffice, and soaked maize may be substituted for compound nuts. In winter the supplementary foods may be largely potatoes or fodder beet, in which case meal or nuts with a rather high protein-content should be used. If sows are kept indoors during pregnancy, with only limited amounts of greens tuff, it may be well to incorporate a trace of iron salts in their food, with the object of ensuring that their young will be born with good reserves of iron and will, thus, be unlikely to develop amemia. Five to 10 gm. (say 1/4 oz.) per head per day is sufficient.

Sows will ordinarily have been served before they join the group but, of course, may return to service; the group should, therefore, be looked over daily for signs of oestrus.

Each female, some ten or fourteen days before she is due to pig, should be moved to her farrowing quarters in order that she may be appropriately fed and have time to become accustomed to her surroundings. The udder should be washed with soap and water to get rid of any worm eggs that may be present on the skin. Her ration should be about 7 lb. of meal equivalent, mostly in concentrate form, rather laxative, and moderately rich in protein. Weatings (middlings or sharps), with the addition of perhaps 5 per cent, of fish meal and 5 per cent, of linseed meal, is a popular diet. An alternative is an ordinary sow-and-weaner mixture, with the addition of perhaps 20 per cent, of bran. Only a small amount of green food should be given. The pen should be kept clean and well-littered with short straw, and the animal should be occasionally handled and petted.

Milk ordinarily appears in the teats about a day before the onset of birth pains, and, soon after, the sow will be seen to gather litter with her mouth and make a nest. Any long straw should now be removed and replaced with clean cavings or short straw; baby pigs arc liable to become entangled in long litter and may fail to reach the teats, or be overlaid.

Nobody except the regular attendant should be present at farrowing, and mere should be no interference with nature except in the very rare cases where help is clearly needed. In many cases, once parturition has begun and is seen to be proceeding normally, the sow can be left for intervals of an hour or so.

The chief danger is from an excited or nervous sow savaging or even eating her young or treading or lying on them, the risk being greatest with the first litter. If trouble is feared, the piglets, as they are born, may be removed to a warm box or barrel and kept covered up until farrowing has been completed and the dam has settled down. Sows that have once killed a litter should not be bred from again. The piglings' mouths should be examined and, if the incisors are very prominent or if the sow is evidently suffering pain during suckling, the teeth should be nipped off with pliers. It is an advantage to leave a light over the farrowing pen for a few nights.

During the first two days after farrowing the sow will take little food and should be offered only a thin slop of middlings or if she is in very weak condition-some warm milk Her demands will remain small for a further few days and will generally be met by a daily ration of 6 or 7 lb. meal, fed as a slop, and divided into two feeds. Many breeders withdraw green food entirely during the first week lest the sow's milk cause some digestive disturbance in her young. By the end of the first week, the flow of milk will have greatly increased, and it ordinarily continues to rise during the succeeding fortnight. The ration must, therefore, be raised to 10, 12, or even, in the case of large sows with large litters, 14 lb. of meal per day (in three feeds) with a small allowance of greenstuff. A rough rule is some 4 lb. for the sow and I lb. for each piglet, but few sows will consume more than 14 lb. The concentrate should be well balanced (see p. 248).

During the second and third week the attendant should watch for symptoms of amemia since, even if a sod is placed in the sty daily, it may be neglected. If the skin and, especially, the ears develop a yellowish instead of the natural pinkish tinge, and if there is any tendency to scour, iron salts should be administered daily until the piglets are seen to be eating or nosing at the sod (see p. 247). The easiest means of prevention is to let the piglets out to grass for an hour or two daily.

To ensure continuing rapid growth after the sow's milk yield begins to drop, as it commonly does during the fourth week from farrowing, the piglets should be induced to eat as soon as may be. They must be encouraged with palatable and nutritious food which, since it must be constantly accessible to them, must be out of reach of the sow. Wet mash, though palatable when fresh, quickly turns sour, and dry meal is apt to be wasted. Thus, small nuts or pellets are the best form. Some pigmen, if interest does not develop, sprinkle the food with milk. Appetite should develop fast and, by weaning time, consumption should have reached fully 1 lb. per day.

Males not required for breeding should be castrated about the fourth or fifth week of age; if the operation is delayed until near weaning time the animal suffers a double check "Rigs" (cryptorchids) should be fed off before they are five months old since, after this time, their flesh develops the strong flavour of boar's meat The minimum age for weaning is seven weeks, and eight or nine is normal. The weaners should be fed thrice daily and if, as is best, they are left in the sty where they have been born, food may continue to be left in the creep. It is preferable, at weaning time, to remove the sow, rather than the piglings, from the farrowing sty.

Regarding the rate of growth, piglings weighing 2 to 3 lb. at birth should reach an average of at least 30 lb. at weaning unless, indeed, the litter is very numerous, when the two or three smallest, having been on the rearmost teats, may bring the figure down. In Denmark and New Zealand 30 lb. is considered to be a poor achievement; indeed, some New Zealand authorities mention 40 lb. as attainable in practice, but the assumption is that separated milk is available in unlimited' amount.

A 35-lb. average is, however, attainable without milk given stock and good management, well selected separated milk has an even greater effect upon growth rate than can be explained by the amount and high quality of its protein and its high digestibility. An allowance of even half a pint per day produces a marked response.

After weaning the growth rate, starting at about ¾ lb. per day, progressively rises to about 1½ lb. per day at six to seven months (200 lb. live-weight). For a further month or two the rate of gain remains about constant and thereafter declines. Breeding gilts, which should be kept in growing condition but not fattened, will ordinarily make rather less than three-fourths of the weight increases of bacon pigs.

The rate of live-weight gain is less important than the economy of gain (food consumption per pound live-weight increase). Upto a point, however, the two go together; assuming that the maintenance requirement is constant in relation to body weight, and that surplus nutrients are devoted to growth without fattening, a fast growth rate will be associated with a low total food consumption per pound of live-weight gain. On the other hand, the formation of body fat requires much more food energy than the growth of bone, muscle, and vital organs. It will be obvious that over-fattening is doubly undesirable for, on the one hand, the food cost per pound of carcass will be enhanced, while the carcass will be valued at a lower price.

Different authors have expressed rather widely different views about the size of carcass that can be produced at the lowest cost per pound. The cost of the weaner, per pound of carcass weight, will obviously be high, since the cost of maintaining a sow for six months, as well as depreciation, must be divided over the litter. In the early stages after weaning the economy of gain is very high, and it progressively falls with age. On the other hand, older pigs can consume increasing amounts of relatively cheap food-waste potatoes, fodder beet, swill, etc. The overall result is that there will often be little difference in costs, per pound of carcass weight, within a range of 200 to 250 lb. live-weight. Porkers of half this size are substantially more expensive to produce, per unit of carcass weight; one calculation makes the difference about 25 per cent.

With regard to the level of feeding there is general agreement that animals being reared for slaughter should be fed concentrates to appetite-as much as they will clean up in about twenty minutes-until the age of nearly six months (by which time the consumption will be about 6 lb. per head per day), but that, thereafter, the allowance should be restricted to about 6 or 6½ lb. daily. In the early stages full consumption should be encouraged either by providing three meals daily or, alternatively, by feeding twice daily and allowing access to a self-feeder. Feeding to appetite after the age of six months will, in most cases, result in an overfat carcass and a poor ratio between food consumption and live-weight gain. The exception may be made in the case of animals of the largest and latest-maturing types, which may profitably be fed to appetite throughout. The poorest carcasses, and the poorest returns for food consumed, result from underfeeding in the early stages (which restricts growth and the development of muscle) followed by over-feeding in the latter part of the animal's life, which leads to a large accumulation of expensive and unwanted fat.

In the rearing of breeding gilts the objective is to secure nearly full growth with very little fattening. This can best be achieved by reducing the allowance of concentrates to about three-fourths of the amount that would be eaten by, or would be appropriate to, a slaughter pig of the same age, and by allowing the animals to satisfy their hunger by grazing, or by consuming such bulky foods as kale, clover, roots, etc., to appetite.

The control of diseases and parasites is of prime importance. The most disastrous of the commoner infections is Swine Fever, and unless there is a high degree of isolation from possible sources of infection, vaccination should be carried out as a routine. The incidence of Swine Erysipelas varies much as between districts, and veterinary advice should be sought on the desirability of immunization. Outbreaks of coughing may be due to virus pneumonia, and veterinary advice should be sought.

Periodic worming is desirable even if there is no obvious sign of infestation, since it is much cheaper to prevent a build-up than to deal with one after it has occurred. The best of known medicaments is sodium fluoride, its only disadvantage being that the dosage must be accurately adjusted to the size of the animals, varying from 1/7 oz. at weaning time to 5/6 oz. at maturity.

A watch should be kept for the appearance of lice, which generally first show about the ears. A soapy emulsion of derris, applied to the skin with a brush, will kill the living specimens but not the nits, so that treatment should be carried out twice with an interval of a fortnight.

The droppings should be watched. If the animals are constipated a mild laxative should be administered by way of the food, and modification of the ration-e.g. the addition of some bran-may be considered. Scouring may be a symptom of serious disease ; on the other hand, it is often due to particular foods-e.g. unripe fodder beet. If the trouble appears to be caused by some irritant in the food, castor oil may be administered, and the animal put, for a few days, on an invalid diet of flaked maize and water.

Outbreaks of fighting are liable to occur, sometimes without obvious cause, but most often when lots of well-grown animals are re-assorted. It may happen that one particular animal is set upon by all its pen-mates and be seriously injured, or even killed. In general, the introduction of strangers should be avoided; it is, indeed, rarely practicable to keep each litter separate from start to finish, and, in fact, weanlings rarely give much trouble. At more advanced age resentment of strangers is much stronger. Sometimes the trouble starts when the pigs outgrow their troughroom, and needs in this respect should be borne in mind when the pens are being filled. In large herds it is possible to work to a plan that both economizes space and conduces to peace in the herd. This 'is to run the weaners, in batches of thirty or forty, in yards and, when they are perhaps four months old, to divide the batch into evenly sized groups of eight or ten for transfer to the fattening pens.

CHAPTER VIII

FARM POULTRY

THE domestic fowl appears to be descended from *Gallus bankiva,* a species of wild fowl indigenous to that region of Asia comprising Burma, parts of India, Sumatra, etc. By the beginning of the Christian era the birds had probably reached Europe, their domestication and spread having been due partly to their value as food producers but also very largely because of the sport provided by cock-fighting. Fowls have proved themselves to be particularly adaptable; they have been moulded into a large number of breeds and varieties, each possessing special characteristics.

In order at the outset to get a broad view of the management of poultry, it may be helpful to compare the life of the wild bird under natural conditions with that of modern fowls kept for commercial purposes. The wild Jungle Fowl breeds only once a year, lays a clutch of eggs, broods and incubates the eggs, and looks after the young for as long as is necessary. The young reach sexual maturity at a year old. Moulting takes place once a year or oftener.

Breeds used in poultry farming have been selected so that the short natural laying period has been extended to a period of perhaps ten months, and the natural clutch of a dozen or more eggs has been increased in many cases to over two hundred. To get this high production, birds have been so bred that they seldom go broody and do not pause in laying during winter. In order to reduce the costs of maintaining unproductive birds, modern fowls have been developed to reach sexual maturity and commence laying when about six months old. In commercial practice the time of hatching is adjusted so that the young birds lay a large proportion of their eggs during winter, which, because it is the "unnatural" season, is the time of relative scarcity and high prices. But fowls that are producing eggs in the winter season require to be protected by suitable housing and provided with ample and correct food. During winter the production of eggs may be stimulated by providing artificial light.

Although the laying powers of fowls have been developed to an extraordinary degree, production does not continue indefinitely, and even where broodiness has been eliminated the time comes when the birds stop laying and go into a moult. There are, however, indications that it might be possible to breed non-moulting strains. In order to get supplies of eggs at all seasons, chickens are normally reared over a three- or four-months period so that the flock as a whole will have fowls commencing and finishing laying at different times. In addition, an endeavour is made to have the young birds start laying before the older ones begin to moult. In spite of much being done to equalize the distribution of egg-production throughout the year, most eggs are laid during the "natural" laying season in spring, particularly in the less intensively managed flocks. In the past, production in March-April-May has been sufficiently high for eggs to sell in a free market at about half the price that obtains in October-November-December; but this relationship has been changed by the increase in the proportion of flocks kept under the battery and other indoor systems.

Egg-production is the main business of most poultry farmers, and in many cases the birds sold for table purposes are surplus stock, of all kinds and qualities, which have had no special preparation. In many districts the price of table birds is too low to justify the special feeding and finishing required for the production of the highest quality carcasses. It may at times pay to finish fowls of the dual-purpose and table breeds, but the light "egg-laying" birds are difficult to sell, and where there is no market for petit poussins it may be best to destroy cockerels of Leghorn and other light breeds as soon as their sex can be determined.

Before going on to the details of poultry husbandry it is worth alluding in a very general way to the management of a commercial flock. In a flock of dual-purpose fowls mating usually begins in December; the first hatchings take place in January and other hatchings follow until April. The chickens are first kept in brooders with artificial heat and then in cold brooders. When the sexes can be recognized the male birds are removed to be reared for the table or otherwise disposed off. The female chicks are brought to maturity so that they will begin to lay when six to seven months old. The pullets provide winter eggs and should continue to lay until they go into moult between the following August and November, when their place is taken by that year's pullets. Considerable losses, averaging occur during the first twelve months of production. Once moulting starts, the hens have usually to be kept until December or January before they lay again, and for this reason, and because a hen in her second year gives but 70 or 80 per cent, of her first year's production, only the best of the birds are kept for a second season. The culls are disposed off as soon as, or even before, they stop their first period of laying. The breeding pens are made up of birds commencing their second laying year, and when this second period of production is completed the two-years-old birds are sold. Only very valuable breeding stock is kept beyond the second season.

Breeds : The breeds of poultry are so numerous that they cannot be considered in detail in this work. Broadly, however, they may be grouped into those that have been developed primarily for egg-production, those that have been selected for table, and the dual-purpose varieties that are suitable for both purposes.

The egg-laying breeds, as the name implies, are all heavy layers. In them the tendency to go broody has been largely eliminated. They have light bodies and are consequently rather interior meat producers; indeed, surplus male birds are usually not worth feeding for table purposes. Being light, lean, and often not very heavily feathered, they are not as a rule hardy enough for the most exposed situations. On the other hand, they are very active and keep themselves fit in the relatively close confinement of intensive houses and backyard runs. They are rather too nervous, however, for keeping in individual cages, and they become very excited when they are moved in folding units. They lay eggs of white colour. Of the breeds in this group the most important is the White Leghom-an Italian variety which was imported to this country by way of America. A typical cock weighs about 6 lb. and a hen 4 ½ lb. Other representatives of this group are the Brown and Black Leghorns, the Ancona, the Andalusian, and the Minorca.

The table breeds show considerable variation in size, egglaying capacity, colour of flesh, colour of egg, and so on. The quality they have in common is heavy fleshing. Some are coarse-boned and slow to mature. The most popular is the Light Sussex, which is of English origin, and combines fine meat quality with very good egg-production. A male bird weighs about 9 lb. and a hen 7 lb. Other table breeds are the Faverolle, the Dorking, and the Indian Game.

The best of the dual-purpose breeds have good carcasses and also produce eggs in such numbers as to rival the specialized egg-laying varieties. They are particularly hardy and good winter layers, which may be because they carry more fat and feather. They have the disadvantage of being prone to broodiness, and unless they are watched and treated their egg yield may suffer on this account. They do not all have the activity that is necessary in birds that have to be kept in confinement, where inactive fowls tend to become fat and excessively broody. They nearly all lay brown or tinted eggs. Rhode Island Reds and Light Sussex are the most popular of the dual-purpose breeds; they are approximately of the same size, the males weighing about 8 ½ lb. and the hens 6 ½ lb. Other dual-purpose breeds are Barnevelders, Orpingtons, White Wyandottes, and Plymouth Rocks.

Selection and Breeding : Under intensive and semi-intensive systems of management, farmers should aim at an average of considerably more than 150 eggs per bird in its first laying year. It is only under conditions where fowls are kept purely as scavengers, at the least possible cost, that low egg production can be profitable. Not only is it essential to breed only from good strains but it is also necessary to rid the commercial flock of birds that are unproductive. In every batch of pullets that is reared, there will be some that are quite incapable of giving a sufficient return: to pay for their keep, and these must be culled and sold for what they will fetch before they occasion further loss. Trap nests can be used to obtain a measure of production, but a very good idea of a bird's laying capacity can be had from its appearance and behaviour. The best hens are those that mature early and lay well in the October-November-December season; indeed it is a good plan to mark pullets that are laying well at this time so that when they are mature they may be retained for breeding. Good layers show an active nervous temperament and search long and far for food, while less desirable birds tend to be sluggish. In good fowls the head tends to be broad yet refined and the eye is bright and prominent; in poorer specimens the head may be too fine or too coarse and "beefy," and the eye dull and sunken. The comb and wattles become large, red, hot, and smooth in birds that are laying, but bluish, shrunken, scurfy, and cold when no eggs are being produced. The bodies of good fowls show roominess and a large digestive capacity. Flexible pelvic bones placed wide apart, and a large moist vent, are other indications that the bird is laying. Fowls which ordinarily possess yellow pigment in the beak, legs, and other parts lose this colour when they are producing eggs continuously; the colour of the beak may be lost after five or more weeks' lay, and that of the legs after fifteen weeks or more. The best birds moult late in autumn- that is, in October and November; they lay right upto the beginning of their moult and then moult rapidly. The September moulters should be culled; but birds may be retained if they have completed their moult by the 1st of September and are again about to lay. When pullets moult in their first year it is generally due to faulty management.

The inheritance of high egg-production is by no means simple: it is probable that separate hereditary factors for early sexual maturity, broodiness, intensity of production, tendency for a winter pause in laying, and persistency in laying, all contribute to the total result. Factors for vigour and digestive capacity add further complications. It is doubtfull if mass selection can really lead to much further improvement; indeed it has been stated that analyses of records show no relationship between the production of daughters and that of their dams or their sire's dams. At best the correlation between a dam's production and a male bird's worth seems to be low. It, therefore, follows that the only effective way of increasing the egg-production of a flock is to use birds that have been selected by means of a progeny test. By contrast with other types of livestock, progeny testing can be applied to females as well as to males. An individual hen can produce enough daughters in a single year to provide good evidence of her genetic make-up. Of course birds must be selected for other things as well as egg yield: the shape and size of the eggs are important matters; the fowls must also conform to breed standards. In a few cases good producers are unsatisfactory breeders because their eggs are of low hatchability.

In the practical mating of the birds the males must be put into the pens some time before the eggs are taken for hatching. Actually the eggs should be fully fertile five days after' the first mating. With dual-purpose breeds a cock is run with 8 to 12 hens, and with light breeds the ratio can be 1 to 10 or 15, but rather more hens may be given where there are several males running in the flock. The use of cockerels produces the greatest proportion of fertile eggs, but a bird that has proved his worth in a progeny test is, of course, kept as long as he is useful. Hens in their second laying season lay larger eggs than pullets, and are generally preferred for breeding since they have proved their ability to survive. Pullets are sometimes used as breeders, but when this is done they should be hatched early, and after laying for a time it is better that they should moult and rest for a while before the start of the breeding season. Fertile eggs will continue to be laid for five days, after the removal of the cock, and by the tenth day fertility will have fallen to about 50 per cent. One reason for low fertility in practice is that some hens may be neglected by the male bird.

Pen mating is very common-that is, each male bird and his group of hens are penned separately. Mass or flock mating is sometimes practised; this consists of running all the male birds with the entire flock of hens. The obvious disadvantage of mass mating is that it is quite impossible to keep any record of pedigrees, but on the other hand, eggs from such mating show high fertility. A technique for artificial insemination has been developed and this enables a very large number of females to be fertilized by one male.

With special breeding stock very elaborate records have to be kept. Egg-production is recorded by means of trap nesting, the individual birds being distinguished by numbered leg-bands. Hatching eggs may be marked with their pedigree and incubated in special bags or cages to keep the chicks separate until they are numbered by leg-or wing-bands. In addition, coloured rings are slipped on the legs of the birds for special purposes. Where trap nesting is not practised the early layers among the pullets may be given a coloured band, so that they can be kept for breeding in the following year.

Sex-linkage in Breeding .-It has already been pointed out that male chickens of certain breeds are often worthless. With other breeds it may pay to rear for table purposes chickens that have already been kept long enough to render their sex apparent, yet if the sex could be distinguished at the outset it might be better to get rid of the cockerels than to feed them for meat. With the older pure breeds there is nothing to indicate the sexes until the secondary sexual characters begin to develop, but by crossing birds possessing certain characteristics it is possible to separate the sexes with almost absolute certainty at the time of hatching. Owing to sex-linkage, the sons of hens having the following characteristics resemble their dams, when got by cocks of the opposed type; conversely, the daughters resemble their sires:

1. A ground colour of creamy silver as opposed to buff or golden brown. For example, Light Sussex, Silver Wyandotte, and Silver Hamburgh hens mated to Brown Leghorn, Buff Orpington, or Rhode Island Red cocks will produce silver males and gold-coloured females.
2. Barred plumage as opposed to unbarred. For example, Plymouth Rock, Scotch Grey, and Cuckoo Leghorn hens mated to Black Leghorn or Black Minorca cocks produce male chicks with a light head patch and black pullets.
3. Light-shanked as opposed to dark (slaty) shanked. For example, White Leghorn and White Wyandotte hens mated to a La Bresse cock give light-shanked male chicks and dark-shanked pullets. These characters, however, are not so easily distinguishable as those in cases 1 and 2.
4. Slow-feathering as opposed to rapid-feathering. For example, hens of the heavy breeds, which tend to be slow-feathering, as opposed to a rapid-feathering breed, such as a Leghorn. Thus, slow-feathering male chicks can be got by using a Rhode Island Red hen and a Leghorn cock. The disadvantage here is that the unwanted birds may have to be kept for some days and even then it may require experience to distinguish the sexes.

It may be of interest to consider the mechanism, whereby this sex-linkage takes place. The cock has two sex chromosomes but the hen has only one; the result is that while every male gamete (sperm) has a sex chromosome, only half the female gametes (ova) are so equipped. Now if the character for colour is carried in the sex-chromosomes, half the ova will carry no colour factor; if one such ovum should fuse with a male cell the offspring can only inherit the colour of the male, and as it has only one sex chromosome it must be female. If the colour carried in those ova which do have sex chromosomes should be dominant to that carried by the male cells, the offspring will have the colour of the mother; and because they get a sex chromosome from each parent they will be males.

Sex-linked breeding can only be carried out where the dominant factor is possessed by the female. For example, if a pure barred cock is crossed with a black hen the chickens of both sexes will be barred. With certain combinations of colour factors, however, even pure-bred chicks are distinguishable at birth. New breeds, such as the Cambar and Legbar have this characteristic. The Legbar is simply a Brown Leghorn to which the barred character (alternating light and dark marks across the feathers) has been added. Legbar pullets can be produced cheaply and easily from any Brown Leghorn stock by mating a pure Legbar male with ordinary Brown Leghorn females. As the barred character is sex-linked it is passed from the father to his daughters. The sons of this cross are also barred, but as they contain their mother's unbarred character they are impure and unfit for breeding; they are indistinguishable from females at hatching and have to be kept until their combs and tails disclose their sex. When pure Legbars are mated together the sex of the offspring is distinguished by the shade and pattern of the chicken down. Unfortunately the progeny of some Brown Leghorn stocks show differences that are not very sharp. Many people can be trained to distinguish the sexes of young chicks by examination of the vent but a considerable amount of practice is necessary before anything approaching certainty can be reached. Chick-sexing instruments, which simplify the process, have recently been developed.

While first-cross hens may often show vigour and productiveness superior to those of their parents, the great disadvantage of sex-linked or other crossing is that the birds cannot be used for breeding because second crosses are of inferior quality.

Incubation : Hatching chickens in the natural way by means of broody hens is too laborious for modem conditions, and about nine birds out of every ten are now hatched by artificial means. Incubators vary from a 50-egg size to mammoth machines which can hold thousands of eggs at a time. While they differ in capacity and in details of construction, they are all insulated containers equipped with one or more drawers to hold the eggs and provided with a heating device and some means of securing a constant temperature. The popular small incubator of, say, 150-egg size (see Plate 32, *A)* has a paraffin lamp and a thermostatically controlled heat-regulating damper; it is constructed so that the egg temperature is maintained by a current of hot air; it may or may not have a quick-turning device fitted to the egg tray. Mammoth incubators are heated by oil, coal, or electricity, and a fan is employed to ensure thorough circulation of the heated air and to maintain a uniform temperature; their numerous drawers can be tilted so as to turn the eggs. Full particulars of the construction, care, and maintenance of incubators can be had from makers' catalogues and handbooks.

Incubators should be housed in a soundly constructed and properly insulated building where there are no draughts and where a fairly uniform temperature can be maintained. A room or a cellar in a dwelling-house may accommodate the incubators for a small farm flock. A hygrometer is a useful fitting in the incubator room, for it will enable measures to be taken to adjust the humidity.

In deciding on the incubator capacity for a particular size of flock there is much to be said for being able to hatch all the required chickens in two batches, and at most it should be possible to do the hatching in three. For hatching on a large scale it is better to use a mammoth machine than a number of small ones.

Of every 100 hens in a farm flock, roughly 75 should be pullets and the remainder yearling birds that have been kept for breeding. In order to obtain 75 selected pullets, about 85 will have to be raised, so as to allow for me culling of 10, and in order to raise 85 it will be necessary, because of some mortality, to hatch 100. As half of the chickens will be males, the hatch will have to be one of zoo birds. But of the eggs put in the incubator it is reasonable to expect that only 60 or 70 per cent, will hatch, so that the total number of eggs required will be 300, or 4 eggs for every pullet that is wanted. Obviously the incubation could be performed by setting a 150-egg incubator twice.

It has already been indicated tot the first hatchings of heavy breeds will take place in January or February; light birds, such as Leghorns, which take about a month less to mature, can be hatched among the last batches. Early hatched chickens do better than late, and in addition, table cockerels from early lots can be marketed to advantage before supplies become too heavy. Hatching for egg-production should be completed by the middle of April. Birds for table purposes may be hatched in autumn so that they can be marketed during the high-price period in early spring. Summer hatching may be practised with a view to the Christmas market, and where a contract can be made for regular weekly deliveries of table birds, hatching may continue throughout the year.

The eggs from the breeding pens hatch best when they are not more than five days old; there is a loss of about 20 per cent, in eggs that have been kept a fortnight, and those over three weeks old may be considered useless for hatching. If eggs have to be collected over a considerable period they should be stored at about 50° F; and in no case should they be kept at a high temperature, which would reduce the vitality of the germs. The eggs should be free from cracks and they must not be washed, but may be dry-cleaned with wire wool.

The incubator is run for two days in order to get it regulated before the eggs are placed in the trays. The temperature is of very great importance and it must be watched with great care. When the thermometer bulb is about I in. above the eggs the temperature should usually be 120°, 103°, and 104° F in the first second, and third weeks respectively; but it is generally best to adhere closely to the makers' instructions regarding this point.

It is necessary to turn the eggs frequently in order to prevent the embryos sticking to the shell membrane. When the turning has to be done by hand it should be performed twice daily from the beginning upto and including the eighteenth day. But eggs benefit from frequent turning, and with machines fitted with quick turning drawers it is worth carrying out the operation four times a day. It used to be customary to cool the eggs periodically, but investigations have shown that cooling in itself does no good. In incubators where the ventilation is poor, however, it may help the hatch if the eggs are aired for five, ten, and fifteen minutes daily in the first, second, and third weeks respectively. Cooling is quite unnecessary in mammoth incubators where the air is circulated mechanically.

In small incubators eggs are tested twice during the incubation period to reveal infertile and dead embryos, the examination being carried out by holding the eggs against the light of a testing lamp. The first test takes place from the fifth to the eighth day, and all the infertile eggs and those with dead germs are removed; a second test between the fourteenth and eighteenth day permits of the removal of eggs whose embryos have died since the first test. Infertile eggs are clear when lamped. In the first test the fertiles show the living embryos as small dark spots radiating blood-vessels, but in dead embryos the blood settles away from the germs and about the edges of the yolk. In the second test the living embryos fill the eggs well but the dead ones show only partial development. In mammoth machines eggs are tested once only, generally on the eighteenth day.

During incubation relative humidity should be between 55 and 65 per cent; that is to say, the air should be neither too dry nor too wet, for both extremes injure the hatch. Ventilation is very important, and especially towards the end of incubation when respiration is active within the egg. If the percentage of CO_2 should rise too high the hatch will be very seriously diminished.

Hatching takes place after twenty-one days' incubation. Temperature and humidity should both be maintained during the last stage; and since inspection merely reduces both, and so injures the hatch, the incubator should not be opened after the eighteenth day until the hatch is completed. Chickens should be kept in the nursery drawer of the incubator for not less than twenty-four to thirty-six hours before being taken to a brooder; during this time they require no food.

Brooding and Rearing : As soon as the chickens are hatched, they are susceptible to infectious diseases and parasites, and successful rearing is, therefore, as much a matter of sanitation as of nutrition. It will help to keep an outbreak of disease within reasonable bounds if the batches of chicks are not too large, and 350 per lot should not be exceeded. Most important of all, as diseases and worms are almost certain to be present to some extent in old birds, chickens should never mix with adult fowls nor have access to soiled ground. It follows, therefore, that brooding should be done either on clean ground or in a system of housing which protects the birds from all risks of infection. Experience suggests that outdoor rearing, while it results in some degree of parasitic infection, produces birds of greater stamina and disease resisting powers than those that are reared on the intensive systems. Day-old chicks or hatching eggs should be bought only from flocks that are free from pullorum disease. Freedom from this can be obtained by blood-testing the breeding stock and by rejecting the reactors.

Brooder houses should be well lighted, well ventilated yet draught proof, and easy to work in and keep clean. They may be small and portable or large and centralized: in the latter case they are divided into brooding compartments and may have a food store and a central-heating plant. The floor space in brooders should be 50 sq. ft. per 100 chickens, and there should be 7 sq. in. of hover space per chick. When only soiled ground is available the brooders may have sun-porches which enable the birds to get air and direct sunlight; in order to reduce the risk of infection the porches may have wire-mesh floors placed I ft. or more above ground.

The essentials of a brooder are a heating device, a cowl to intercept and hold the heat over the brood, and a control to maintain a set temperature: the actual arrangement under which the birds seek warmth and comfort is usually a hover (see Plate 32, *B);* but a brooder may be simply a ventilated wooden chamber with a heating lamp. The smallest sizes, which are portable, are generally heated with a paraffin lamp and made for batches of about 100 chicks: they can be used one to each brooder house or one to each brooder compartment. Electricity is very suitable for heating brooders in fixed houses, if it can be got at a reasonable price, and it has the advantage that the temperature can be very exactly controlled by an electric thermostat; it saves much trouble and cleaning. Anthracite stove brooders are very good for batches of 200 chicks and upwards, one being used for each house or division. A centralized hot-water system is useful for large brooder houses, the water being taken to a radiator in each brooder compartment, or the hot-water pipes buried in the concrete floor so that the whole or part of the floor itself is heated. Battery brooders consist of a series of wire-floored compartments placed one above the other with a droppings tray under each. The temperature in each compartment is adjusted according to the age of the chicks. The birds in anyone compartment are forced to remain in the same temperature, which has an enervating effect; and although growth may be rapid, feathering is poor and stamina somewhat lacking. Tier or table-type brooders are modified battery brooders, in which there is a cool as well as a heated section in each compartment, so that the chickens can get away from the heat. This arrangement overcomes, to some extent, the limitations of the battery brooder, and tier brooders are quite suitable for rearing commercial stock for the first few weeks. For rearing potential breeding stock less intensive methods are to be preferred.

Brooders should be got going a couple of days before the hatch is expected, so that the temperatures can be adjusted and the surroundings dried and warmed. The chicks should be brought in a container which will prevent their being chilled. A ring of netting is first placed round the brooder to keep the birds from wandering too far from the heat; it is widened from time to time until it is taken away after about four days. The temperature under the hover should be about 90° to 95° F. to begin with, and it should be reduced gradually to 70° or 75° F. by the end of the fourth week, after which the heat is kept constant until it is needed only at nights; by the sixth or ninth week, depending on the season, it may be discontinued altogether. Actually, however, the temperature is adjusted according to the behaviour of the chicks: if too cold the birds crowd, while if they are too hot they gasp and show distress, or move away from the heat. In very cold weather heat must be supplied until there is no more risk of crowding. To make sure tot ventilation is adequate, the curtain of the hover should not quite reach the floor, and the hover is gradually raised as the birds grow and harden off. Roosts are introduced at about four weeks so as to encourage the birds to perch, and the sexes are separated at six or eight weeks.

The unwanted cockerels are transferred to pens for fattening. The pullets should be put on clean ground-in slatted-floor arks, folds, or colony houses-and there be retained without further change until they are ready for the laying house. The perching requirement in the house will be 6 in. a pullet by the time the birds are well grown. After a change over, the fowls are kept housed until they are used to their surroundings, or a net can be run round the house to keep the birds from straying too far. The stock cockerels, like the pullets, should be raised in clean, healthy surroundings .

Housing : Apart from the incubator room and storage for utensils, food, tools, etc., houses are required for *(a)* Brooding, *(b)* Rearing, *(c)* Laying, and *(d)* Breeding. The actual selection of houses must depend on the size of the flock, the ground available, the system of management, and so on. Before dealing with the systems of poultry keeping it may be helpful briefly to review the types of house[1] that can be employed for the different flock groups.

Brooder Houses : (a) Portable brooders or "nurseries" made to take about 100 chicks: they have a well-insulated chamber that is heated by a lamp, and a cooler chamber where the birds normally exercise and feed; they should be placed on clean grass, so that the chickens may be allowed out during good weather and when they aire older, *(b)* General-purpose

[1] For the details of house construction, see *The Housing of Poultry,* published by the Ministry of Agriculture.

houses, each fitted with a portable hover or indoor rearer, and capable of taking from 15° to 25° chicks. They can at a later stage be fitted with perches and so serve for rearing; indeed they may be used for pullets right through their laying period (see Plate 33, *A), (c)* Multiple-unit brooder houses, provided with a passage-way and divided into compartments, each of which is fitted with a hover, *(d)* Battery-brooder houses. These have to be specially insulated and ventilated to deal with the large-capacity brooders.

Rearing Houses : (a) General-purpose houses can, as stated above, be used for rearing as well as brooding, *(b)* Slatted-floor arks or folds, which are easily moved, may be used when the chicks are ready to leave the brooder. Where free range is not given, each ark may have a netted run, *(c)* Range shelters may be constructed very cheaply and serve for growing birds. They are generally made with a sound roof over the perches, and the wall frames are merely covered with netting.

Laying Houses.-(a) General-purpose houses of the type used for brooding and rearing can, as already stated, be used for laying. When mounted on wheels or runners the smaller sizes can be moved, with the aid of a horse, from one part of the farm to another. These houses are designed for the semi-intensive system, are provided with netted runs, and are sited on grass. An alternative is the henyard system, under which the runs may be considerably reduced in size and kept littered with straw. In the former case the house can be used to confine the birds in bad weather, but if small enough to be conveniently portable it will only hold 12 to 20 hens. In the latter case a portable, house with 40 sq. ft. of slatted floor will accommodate 40 to 50 laying birds, and this results in relatively cheap housing; but the hens cannot be confined, *(b)* Fixed laying houses to take 100 to 150 birds. Since the best yields are usually obtained from fowls running in small groups, houses of greater capacity should be divided into 'sections accommodating not more than 150 birds each. These houses are designed for the semi-intensive system of poultry husbandry and they are provided with netted runs, *(c)* Intensive houses where birds are confined all the time and which, for the sake of cheapness in construction, may be several stories high; they may have sun porches, *(d)* Folding units with slatted floors and attached out-runs (see Plate 33, *B)*. These accommodate about 25 birds and are provided with some device whereby they can be moved by hand daily on to clean land, *(e)* Laying battery houses where each individual hen is placed in a wire-fronted cage, the cages being arranged in several tiers one above the other.

Breeding Houses : (a) The general-purpose colony house can be used for flock mating, *(b)* Houses to hold a male bird and his complement of hens may be set in separate enclosures, *(c)* Folding units can also be used as breeding pens.

Male birds can be reared in general-purpose houses, arks, or range shelters. The fattening cockerels may be kept in spare buildings about the farm and finished in fattening coops.

The houses have to be equipped with water fountains, mash hoppers, grit boxes, etc., of a size suitable for the age of the birds. Nests and broody-coops are, of course, needed inlaying houses. The hoppers should be so constructed that the birds can neither scratch out nor foul the mash.

Systems of Poultry Keeping : *The Free Range System* consists of running the fowls on unfenced ground. It is suitable for a general farm because, if houses are provided to accommodate about 50 birds per acre, and placed at least 80 yds. apart, the ordinary stocking of the land will suffer little interference and the birds will remain in their own colonies. Movable general-purpose houses are very suitable for this system, as they can be pulled on to other ground, when the land is to be broken up for cropping, or they may be run on to the stubbles after harvest. Even when the houses are fixed, the light stocking ensures that the ground will not be seriously contaminated. As no fences are required, and as slatted-floor houses holding a laying bird per square foot of floor may be used, the cost of housing the flock is very small; moreover, there is a further saving because the wide range provides the fowls with all their greenstuff and much insect and other animal food. On the other hand, the unprotected chickens and hens may suffer severe losses from birds and beasts of prey, and the wide spacing increases the labour of attending to the flock.

If general-purpose houses are employed they may be used in conjunction with portable hovers for brooding the chickens, and later on for rearing, laying, and flock mating. Quite often, however, the brooding arrangements are centralized and the free range is used only for rearing and laying.

The Folding System consists of keeping the fowls in small portable houses with attached wire-netting out-run, the whole forming a "folding unit." The houses have slatted floors and are provided with all that is necessary in the way of nests and food- and water-containers; they are usually made to hold two or three dozen hens. The units are moved daily on to fresh ground so that the birds, although relatively closely confined, remain healthy; the droppings are evenly spread and improve the herbage. As the run is completely netted in, there is little danger from vermin. This system is very suitable for a mixed farm and can be used on pasture that is carrying other stock. On the other hand, the method is being used fully to stock the land at the rate of about 400 birds per acre. The units are suit able for rearing birds and for layers, and they have even been used for breeders; they are unsuitable for brooding young chicks.

The Semi-intensive System is one in which a considerable degree of centralization is obtained by providing fixed houses and permanent runs. Wire netting is used to make the enclosures, and it is common to provide two or more out-runs for each house, so that while one is in use the other may be rested or even dug over and re-sown. The stocking should not exceed 300 hens per acre. The system is economical in labour and in the use of the ground but it requires extra capital for fencing, and the birds can hardly be expected to do so well on old runs as on fresh young pastures. The floors are littered with straw, chaff, or other suitable material, and provide scratching and exercising ground, when the birds are confined during spells of very bad weather.

The henyard is a modification of the semi-intensive system. In place of the grass run the birds are given access to an uncovered but well-sheltered yard which is periodically littered with straw. The house should provide 2 sq. ft. per bird, and the yard about twice as much.

In this system brooding can generally be carried out most economically in a multiple-unit brooder house. The young birds may be reared in the ordinary houses and runs, but it is a great advantage to have clean ground on another part of the farm, where they may be kept in slatted-floor arks or in range shelters. Small houses and runs are usually provided to accommodate the breeding fowls, and these permit of pen mating.

The Intensive System is one wherein the laying stock are kept entirely in confinement. The houses should be well insulated and ventilated. Ample light must be provided. Each bird is allowed about 3 to 4 sq. ft. of floor.

Formerly the floor litter was removed periodically, but now it is commonly left to accumulate. If environmental conditions are satisfactory, composting of the litter and droppings occurs, heat develops, and the litter remains dry and friable. Periodic additions of fresh litter are made; hence, the method is described as the "built up" or "deep litter" system. It is labour saving and requires less capital to start, but demands a higher degree of skill in feeding and management than the battery system. In the battery system the birds are kept in individual cages arranged in tiers after the fashion of battery brooders. The whole arrangement is called a hen battery, Very little ground is required for this system and it is relatively labour saving, especially at week-ends. As individual egg records can be kept, culling is simplified; and battery culls command good prices as table birds, for their flesh is obviously more tender than that of birds allowed free range. On the other hand skillful feeding is necessary, as the birds have no opportunity of getting natural food and so correcting deficiencies in their rations. Birds kept intensively for commercial egg production are generally disposed of after one laying season.

During the short winter days egg production can be increased by using artificial light to lengthen the day to one of about fourteen hours. This extra light, functioning through the eye of the bird, stimulates the ovary to produce more eggs, and consequently the bird requires more food. The old theory that additional light increased egg production simply by allowing the bird more time to eat is no longer held. Once extended lighting has been introduced it must be continued, otherwise the birds will be thrown into a partial moult and cease laying. Artificial lighting, when started in late August, is particularly helpful in preventing pullets hatched late in the previous autumn or in January or February of the same year from going into a premature moult.

Under the head of "Brooding" it has been pointed out that poultry rapidly contaminate the soil with worm ova, coccidia, and other disease organisms; hence, heavy or continuous stocking will sooner or later result in many unthrifty birds and smaller returns. It is particularly important that young chickens and developing pullets should be protected from disease and kept off soil that has carried poultry within recent years. Coccidiosis has long been the worst of the diseases associated with contaminated ground, but this can now be largely controlled by routine treatment with sulpha drugs, especially sulphamezathine, added to the mash or drinking water at the rate of 0.2 per cent, (about 1 oz. to 3 gals.). The new drugs sulphaquinoxaline and nitrofura-zone are effective both for prevention and cure. With expert knowledge of nutrition, brooding may be carried out altogether in confinement, but there is less risk of failure to rear vigorous and healthy birds if they can have a spell on fresh clean grass. The farmer is much better able to provide clean land than is the specialist poultry keeper. Admitting the advantages of having the birds situated centrally near the homestead, it would nevertheless be best to start the brooding on clean land each season. Using general-purpose houses, the chickens can be reared each year on fresh ground and can be kept on in the same houses as layers and later on as breeding hens. If three sees of houses are in use, the hens will be out of the first ones in time to have them dismantled and re-erected on the new ground for the next brooding season. This may not be the most economical method of housing, and the wide scattering of the birds may make for laborious working, but it is a system which will enable heavy egg-production to be maintained with a minimum of losses. Even where the laying hens are centralized, brooding and rearing should certainly be carried out with the main object of ensuring the health of the stock.

Feeding : Unfortunately, it is not possible to ration fowls on the basis of the energy and protein values of the foodstuffs given in the Appendix. For one thing, foods are not equally digested by hens and ruminants, fibre in particular being of relatively little value to the former. Accurate rationing is also made difficult by the grazing and insect-eating habits of fowls, for it is only under intensive management that they are entirely hand-fed. However, it should be remembered that the hen has a limited digestive capacity and that she must not be given much fibrous food, such as bran, or her appetite will be satisfied before she has ingested sufficient nutrients. On the other hand, she can deal with a considerable amount of food like cooked potato, which is bulky only because of its high content of water. The heavier the production of eggs, the more the birds are likely to eat and the more important it is that the food should be in a concentrated form. Moderate restriction in feeding results in fewer eggs and not in loss of condition, and there are records of a 25 per cent, reduction in the food allowance causing the egg yield to drop to less than half.

Most of the food given to poultry consists of cereal grains and wheat-milling by-products mixed with a certain proportion of protein-rich material. Yellow maize and wheat are the best cereals; oats are poorer only in so far as they have a lower nutritive value; barley is good but not very palatable; rye is unpalatable. Rice and some other grains are satisfactory if they can be got cheap enough. Wheat bran, sharps, and middlings are all used for making up mashes, and about 20 per cent, of total wheat offal should always be included. Bran is very popular because it "lightens" the mash and perhaps because it is a little laxative it is employed to satisfy the appetite where "forcing" is not wanted. The common protein foods are white-fish meal, meat and-bone meal, skim milk and milk powder, and to some extent; soyabean meal and one or two other vegetable foods.

Mineral matter is of great importance in poultry feeding. Young chickens grow so rapidly that they require a great deal of lime and phosphorus for their frames; laying birds require large amounts of lime, as egg shells are almost pure calcium carbonate. It can be realized how big the risk of deficiency really is from the fact that a hen would have to consume 12 lb. of wheat to get enough lime for one egg. Common salt is another essential, but it must be given with care, as in excess it can be poisonous. Bone meal and steamed bone flour are used to supply calcium and phosphate; crushed oyster shell and small limestone chips are used to provide calcium.

Poultry, like other animals, require vitamins in their food. Vitamin A is supplied in plenty if the birds get green food or carefully dried greenstuff, such as clover or lucerne meal. Yellow maize (but not white maize) is another fairly good source of vitamin A. Most of the factors of the vitamin B complex can be got from green food (fresh or dried), milk products, and dried yeast. Vitamin B_{12} which is a part of the "Animal Protein Factor," may be obtained from natural sources, such as droppings, earthworms, and grubs, but usually for breeding stock some animal protein is added to the diet to ensure good hatchability. Vitamin C is not required. Vitamin D must be supplied to young chicks that do not get direct sunlight or ultraviolet light, for without it they develop leg weakness. Green food contains very little of this vitamin, though sun dried lucerne or grass meal is a good source. The usual means of providing a supply is to give tested cod-liver oil at the rate of approximately 1 per cent. of the ration (1 pint per cwt), but other vitamin-rich oils and irradiated yeast are also used. Laying birds do not need cod-liver oil except during the dull, short winter days, or where they are confined for long periods and all their light comes through ordinary glass. Special window materials are obtainable which allow ultraviolet light to 'pass, or open windows screened with sparrow-proof wire netting will serve the same purpose.

When antibiotics, such as penicillin and aureomycin, are included in poultry diets they normally have a stimulating effect on chick growth, but do not improve egg production or hatchability. As less than ¼ oz. of these antibiotics in a pure state' is required per ton, great care must be taken to ensure thorough mixing with the food. Until more is known about the action of antibiotics, they should only be used in the rations of chickens and turkeys intended for the table. Both soluble and insoluble types of grit are available for poultry. Insoluble (flint or gravel) grit enables birds to utilize their food efficiently, and where a natural supply is not available, it should always be provided in addition to soluble (limestone or oyster shell) grit. The latter type of grit supplies the birds with calcium, but cannot be relied on as a grinding agent; and if this grit is fed alone (without insoluble grit), there is the danger that birds will consume excessive amounts in an attempt to obtain grinding material, thereby upsetting their calcium metabolism with adverse effects upon their health.

In rationing the birds, cereals are commonly ground to a coarse meal and thoroughly mixed with other foods; such as bran; meat meal, and salt, to form a " mash." The mash may be fed at intervals in a wet or crumbly condition, but owing to the trouble that this involves it is more common to feed it dry and to leave the birds with a constant supply. Kibbled maize or whole wheat or oats may be scattered on the ground, or preferably among the litter, to provide the morning and evening feeds. Feeding the grain in troughs is more hygienic where disease is prevalent and it has the further advantage of speeding up consumption during winter. Green food may be given fresh or in a dry form in the mash, but when the birds are on range they will obviously look after their own needs.

In chicken feeding the highest quality of materials should be employed and made up as a dry mash. Milk or milk powder has a particular value for the youngest birds. To begin with, the mash is placed on boards and given about five times daily, but as soon as possible it is fed in hoppers. Water or milk is given to drink and green food forms a useful supplement. A common type of mash for young chicks is made up as follows: (1) 50 parts of yellow maize meal; (2) 30 parts of ground wheat or a mixture of bran and middlings; (3) 15 to 20 parts of protein-rich food including bone-say 8 parts of meat-and-bone meal and 8 parts of dried skim milk, *or 2* parts of steamed bone flour, and as much skim milk as the chicks will drink; (4) I part of salt; and (5) 1 to 2 parts of tested cod-liver oil. In addition, fine oyster shell or limestone grit and fine pebble grit should be available in self-feeding hoppers.

From the time that the sexes are separated the protein-rich fraction of the mash should be reduced gradually. By the time the birds are ready to go on range the milk is stopped and the protein supply should be very small; steamed bone flour can be added to maintain the supply of phosphate. A liberal supply of protein was believed until recently to induce early sexual maturity, but it has now been shown that the effect is due not to the protein but to the mineral matter (chiefly salt) associated with proteins of animal origin. When the pullets are in their rearing quarters they may get more than half their food as whole grain, given night and morning. Their feeding should enable them to grow well and accumulate some reserve of fat.

The following table shows the approximate relationship between age, weight, and food consumption in chickens of dual-purpose type that are being reared as layers :

Age in Weeks	**Average Weight of Chick in lb.**	**Average Consumption of Dry Food per Day in lb.**
0	0.08	...
1	0.11	0.02
4	0.4	0.05
8	1.2	0.11
12	2.2	0.17
16	3.0	0.20
20	3.7	0.24
24	4.3	0.25

Laying birds with some range on grass consume on the average about 4 oz. of food per day; birds in intensive houses may need over 5 oz. Half of this is usually given in the form of a light feed of grain in the morning and a heavy feed of grain at night-the remainder being mash, provided in self-feeding hoppers. If hens in confinement are getting too fat, the protein- and fibre-content of the ration may have to be adjusted so as to reduce the proportion of carbohydrate. Cracked maize, wheat, and oats in various proportions form the grain or scratch ration, while cereal meal, wheat feeds, and fish meal or meat-and-bone meal form the mash ration. Most of the laying rations which have been shown to give good results contain about 10 per cent, of fish or meat meal, but less protein will suffice where the fowls are given an adequate supply of salt and other minerals; indeed, on good grass range there is no need to use a protein supplement, a cereal plus salt ration being all that is necessary. The amount of common salt required, in addition to limestone grit and green food, is about ½ per cent, of the ration.

The water requirement may be reckoned at 1 gal. per day for every 20 birds.

Table Poultry : Cockerels of the light breeds are not worth fattening, but some may be used to supply the limited demand *for petit poussin-birds* of about a pound in weight. Early hatched dual-purpose cockerels can often be reared and sold at a profit, for there is a fair demand for spring chickens from March to June; but later in the season the surplus becomes heavy and prices tend to drop to an unremunerative level. The Christmas demand for high-quality poultry usually justifies the summer hatching of table birds.

Table chickens are frequently reared intensively for the first eight weeks or so and then run outdoors for a further six to eight weeks. They may be marketed directly off the run as " country" chickens; but if a better finish is required, they should be brought indoors to a darkened building for fattening in crates, being fed a slop of Sussex ground oats and skim milk. This feeding period must be short-usually from ten to fourteen days-otherwise appetites flag, but an increase in weight of about ½ lb. per bird is normally obtained. Professional fatteners used to finish the birds by a further period of cramming, but this is no longer economic. Surgical caponization of table cockerels is now seldom practised, but similar results are obtained by treatment with female sex hormones, such as stilbrestrol or hexrestrol.

Intensive table poultry production has been started in England on the lines of American "broiler" production. The "deep litter" system is employed, and at least three large batches of chickens are produced in a year, usually at an age of twelve to fourteen weeks. All the chicks in a batch are hatched at the same time and are reared together in one pen in a large house, allowing about I sq. ft. of floor per bird. These chickens are all marketed at one time, so that the house can be cleaned and rested in readiness for the next batch. The system is economical in labour and equipment. As the birds are kept in one house throughout, they are not upset by the frequent handling and changes of environment generally associated with more conventional methods of table poultry production. Where a highly concentrated, low-fibre, starchy diet, relatively rich in protein and including antibiotics, is fed to suitably bred chickens, birds weighing 4 lb. at twelve weeks of age, which have used less than 3 lb. of food to gain 1 lb. in live-weight, can be produced.

As hygiene is so closely related to poultry husbandry, students are recommended to include the common poultry diseases in their veterinary studies. A bulletin on poultry diseases is published by the Ministry of Agriculture.

DUCKS, GEESE, TURKEYS

Ducks : There are a good many breeds of duck, some of which have been developed for both table and egg-laying purposes while others have been selected primarily for egg production. An example of the former is the Aylesbury-a white duck with a horizontal carriage, a pinkish-white bill, and bright orange legs and feet; it weighs 9 to 10 lb. in the male and 8 to 9 lb. in the female. The Indian Runner is an example of the smaller heavy laying type. There are black, white, and other coloured varieties. The carriage is very erect; the weight is about 4t lb. in the male and 4 lb. in the female.

As compared with hens, ducks have a number of advantages. They are less prone to disease and remain profitable for about three years instead of two; they lay as many and in some cases bigger eggs; the ducklings are ready for market sooner than chickens and yet make heavier weights; they are cheaper to house and can be kept within bounds by quite low netting. On the other hand they have a few disadvantages. Duck eggs are liked by some people and sell readily in certain districts, but because of their rather more distinct flavour they are not so generally acceptable as hen eggs. Another point is that ducks require at any rate 50 per cent, more food than hens. Again, they are much more easily excited, and through nervousness, rendered unthrifty and unproductive.

Ducks can be housed in most kinds of shed or building. They remain on the floor and lay their eggs on the ground, so it is desirable that the floor should be easily cleaned and at any rate the corners and sides should be well-strawed. Special duck houses need be high enough only for convenience in cleaning; a concrete or other impervious floor is an advantage. There should be 2 to 3 sq. ft. or more of space per duck and flocks should not exceed fifty.

Ducks lay batches of 20 to 40 eggs and then pause before starting again. For breeding they are mated in spring-four or five to a drake-and hens are frequently used to incubate the eggs, which take twenty-eight days to hatch. When well-fed the ducklings grow rapidly and are fit to leave the hen soon after they are a fortnight old.

Ducks are fed on the same sorts of food as hens, but the mash is supplied in a much wetter condition and the grain is given in a shallow pan of water. When on free range the birds should be kept slightly short of what they will eat, so that they will forage well and secure the balance for themselves. Plenty of water, grit, and green food are needed, if the range is restricted. Swimming water is unnecessary, but the birds enjoy it. Fattening birds must be kept away from water, otherwise they take too much exercise. Newly hatched ducklings are fed as frequently as possible, but after eight weeks they will do on three meals a day. Adults on range need be fed only twice daily.

Geese are kept primarily for their flesh. The birds feed largely on grass and do best when given an unlimited range of pasture. In practice mash should be fed upto the age of six weeks, and again for a fattening period of about four weeks. During the intervening period, they do well on pasture alone. In spring the numbers are built upto take full advantage of the grazing season; by Christmas the young geese have been marketed and the flocks reduced to the breeding birds. Geese live for a long time and have been known to lay for twenty successive years; they are very hardy.

For table purposes a carcass of 6 to 9 lb. is in demand, and this can best be got from the smaller breeds, such as the Chinese and the Roman, which weigh 10 to 141b. when full grown. Large geese-weighing, according to the breed, 20 or 30 lb. as adults are rather too slow-maturing and coarse for modern requirements.

Geese do not make much use of houses, so these for small numbers can be very simple. Large flocks, however, should be locked up at night in roomy and well-littered houses. For mating, a male is allotted to three females of the heavy breeds and upto five of the smaller sorts. Eggs are laid from February onwards, the number for the season being 20 to 80 according to the breed and the individuality of the bird. The incubation period is thirty days and the eggs are normally hatched by hens, which can cover 4 eggs. An alternative plan is to start the eggs in an incubator and transfer these to hens after a week or two. The young goslings require to be fed at least five times daily, but they soon become independent, and if it is not too early in the season, can do without heat after a fortnight. After about a month three feeds will suffice, and later on, when the birds are grazing hard, one feed per day is enough; indeed "green" goose is fattened on grass alone. However, in autumn, when the pastures are getting bare, birds that are being fattened for the Christmas market should have as much food as they will clean up twice a day. About 3 oz. of concentrated food per day is sufficient for a goose on grass during the breeding season Geese consume the usual poultry mashes and grains, the latter being given in shallow containers of water; they also take chopped greens, boiled potatoes, and kitchen scrap.

Turkeys, like geese, are kept because of the high quality of their flesh, and most of them are reared tor the Christmas trade. The birds are heavy: in the case of the well-known American Mammoth Bronze a mature cock weighs from 30 to 40 lb. and a hen 18 to 26 lb.; cockerels and pullets for table purposes generally weigh 20 to 25 lb. and 14 to 20 lb. respectively. Their breeding life is four to six years.

Turkeys are much less hardy and resistant to disease than are geese, particularly, when they are young. They do best when kept on well-drained soils with abundance of room and the natural shelter of woodlands, and away from ground that has been soiled by other poultry. They require perches but may be housed in rough sheds if the situation is sheltered; they roost readily in trees. The hens tend to lay in places like hedge bottoms, so baskets or boxes-or old barrels on their sides-may be fixed in such sites to serve as nests.

The hens begin to lay in March and produce two clutches of 18 eggs or more. A male is allowed from 10 to 15 hens and one mating fertilizes a full clutch. If only 2 or 3 hens are kept, the male should not be left with them all the time or he will pester them. The eggs from at any rate the first clutch are removed and placed under hens, or hatched in incubators, and the young reared in artificial brooders. The eggs take twenty-eight days to hatch. Sometimes, the turkeys are allowed to incubate their second clutch, but when sitting they are not very amenable to handling.

Adult turkeys are fed on mashes and grains similar to those used for other poultry, and considering their size, they are not heavy eaters. The birds are fond of insects and worms, and when they have a good range, 5 oz. of concentrated food per day will meet their requirements. Young turkeys require mashes with a much higher protein-content, and considerably more vitamin A and vitamin D than those commonly fed to chickens. Newly hatched turkeys tend to have poorly developed instincts and often require training to eat and drink, after the first week they may get five feeds, then four, and from the tenth week onwards three will do. Fattening birds will be of the best quality, if they can be finished on meals and separated milk.

INDEX

Beef cattle 83
  production 76,103,139
  type shorthorn 85
Belgian horse 215
Belted Galloway cattle 91
Berkshire pig 235
Biological value of proteins 40
Biotin 50
Blackface sheep 169
Bloat 126
Blood meal 40